Biomedical Sensing and Imaging

Biomedical Sensing and Imaging

Editors

Wuliang Yin
Yuedong Xie

MDPI • Basel • Beijing • Wuhan • Barcelona • Belgrade • Manchester • Tokyo • Cluj • Tianjin

Editors

Wuliang Yin
University of Manchester
UK

Yuedong Xie
Beihang University
China

Editorial Office
MDPI
St. Alban-Anlage 66
4052 Basel, Switzerland

This is a reprint of articles from the Special Issue published online in the open access journal *Biosensors* (ISSN 2079-6374) (available at: https://www.mdpi.com/journal/biosensors/special_issues/Biomed_Sens_Imaging).

For citation purposes, cite each article independently as indicated on the article page online and as indicated below:

LastName, A.A.; LastName, B.B.; LastName, C.C. Article Title. *Journal Name* **Year**, *Volume Number*, Page Range.

ISBN 978-3-0365-2454-2 (Hbk)
ISBN 978-3-0365-2455-9 (PDF)

Contents

About the Editors

Wuliang Yin was appointed as an MT Sponsored Lecturer at The University of Manchester, U.K., in 2012, and promoted to a Senior Lecturer in 2016 and Reader in 2020. He has authored one book and more than 300 articles, and was granted more than ten patents in the area of electromagnetic sensing and imaging. Mr. Yin was a recipient of the National Scientific and Technical Progress Award by the Chinese Government in 1999, the Excellent Paper Award–Agilent Measurement Forum in 2000, the Williams Award from The Institute of Materials, Minerals and Mining in 2014 and 2015, the 2016 Best Graduate Student Paper from the IEEE I2MTC as a supervisor, the 2018 IEEE Instrumentation and Measurement Society Graduate Fellowship as a Supervisor, the 2020 Best Application Award from the IEEE Instrumentation and Measurement Society, and the IEEE gold medal as the most productive reviewer in the U.K. for IEEE TRANSACTIONS ON INSTRUMENTATION AND MEASUREMENT in 2020.

Yuedong Xie received a Ph.D. degree in Electrical and Electronic Engineering from the University of Manchester, Manchester, U.K, in 2016. After that, he worked as a Research Associate at the University of Manchester and a Research Fellow at the University of Warwick for 3 years. At present, he works as an Associate Professor in Beihang University in China. His research interests include electromagnetic NDT, ultrasonic phased array, electromagnetic acoustic transducers, nano-devices and energy harvesting. He was a recipient of the Best Student Paper Award (First Place) and the Student Travel Grant Award from the IEEE International Instrumentation and Measurement Technology Conference in 2016. He is the author of more than 30 peer-reviewed publications, and he has also given several invited oral presentations in the last five years.

Preface to "Biomedical Sensing and Imaging"

Biomedical sensing and imaging technology are of great significance for the early detection, rapid diagnosis and precise treatment of diseases. In the last three decades, considerable research efforts have been devoted to biomedical sensors and imaging systems. Biomedical sensing and imaging systems can be roughly divided into optical imaging technology, electrical imaging technology, acoustic imaging technology, radiographic imaging technology, etc. As each technology has its own unique advantages, various new types of sensing and monitoring equipment have been invented, and improving the performance of the existing equipment is essential. In recent years, biomedical sensing and imaging have been applied in many aspects of biomedical research, such as biomolecular testing, health monitoring, and disease diagnosis and treatment. Biosensors convert biomedical signals to electrical signals for further acquisition and processing using downstream devices and algorithms, which are also crucial for rapid and accurate biomedical diagnosis. At present, biosensor design emphasizes portability, networked detection, integration and intelligentization. Moreover, research methods are being developed regarding multi-modality and multi-function fusion, which has become a trend in the development of biological imaging systems. Some multi-modal imaging equipment has been successfully used in clinical applications, which can provide more accurate and high-resolution anatomical, functional and biochemical information. With the development of computer technology, physics, molecular biology, materials chemistry and other disciplines, new software and hardware devices have been developed, opening up a new world for the development and application of biological imaging technology. This book, which is a Special Issue of Biomedical Sensing and Imaging, aims to collect recent advances in the above topics regarding biosensing and imaging, especially in wearable/smart biosensors and advanced imaging algorithms.

Wuliang Yin, Yuedong Xie
Editors

biosensors

MDPI

Review

Biomedical Applications of Electromagnetic Detection: A Brief Review

Pu Huang [1], Lijun Xu [2] and Yuedong Xie [1,2,*]

[1] School of Instrumentation and Optoelectronic Engineering, Beihang University, Beijing 100191, China; huang_pu@buaa.edu.cn
[2] Beijing Advanced Innovation Centre for Big Data-Based Precision Medicine, Beihang University, Beijing 100191, China; lijunxu@buaa.edu.cn
* Correspondence: yuedongxie@buaa.edu.cn; Tel.: +86-186-3619-8648

Abstract: This paper presents a review on the biomedical applications of electromagnetic detection in recent years. First of all, the thermal, non-thermal, and cumulative thermal effects of electromagnetic field on organism and their biological mechanisms are introduced. According to the electromagnetic biological theory, the main parameters affecting electromagnetic biological effects are frequency and intensity. This review subsequently makes a brief review about the related biomedical application of electromagnetic detection and biosensors using frequency as a clue, such as health monitoring, food preservation, and disease treatment. In addition, electromagnetic detection in combination with machine learning (ML) technology has been used in clinical diagnosis because of its powerful feature extraction capabilities. Therefore, the relevant research involving the application of ML technology to electromagnetic medical images are summarized. Finally, the future development to electromagnetic detection for biomedical applications are presented.

Keywords: electromagnetic detection and biosensors; electromagnetic biological theory; biomedical application; frequency; machine learning

Citation: Huang, P.; Xu, L.; Xie, Y. Biomedical Applications of Electromagnetic Detection: A Brief Review. *Biosensors* **2021**, *11*, 225. https://doi.org/10.3390/bios11070225

Received: 28 May 2021
Accepted: 3 July 2021
Published: 7 July 2021

1. Introduction

Recently, the biological detection technology has developed rapidly. It can be used to detect the relevant parameters that characterize the performance of the organism due to their unique identification ability to the organism [1]. When the organism is stimulated externally, the organism converts these stimulation signals into electrical signals that can be received and processed. This attribute allows it to obtain nutrients or stay away from danger. Human beings use the sensitivity of biometrics to observe and understand the living environment. In other words, the core of biological detection is converting observable things into measurable physical quantities as the result of biological simulation. In fact, biological detection technology has made greater breakthroughs in bio-electric signal conversion because of the continuous integration and penetration of biotechnology, chemistry, biology, physics, electronics, and information technology [2–4]. According to the detection mechanism, biological detection technology can be divided into electromagnetic biological detection technology, electrochemical biological detection technology, optical biological detection technology, semiconductor biological detection technology, thermal biological detection technology, piezoelectric crystal biological detection technology, etc.

Compared with other detection technology, electromagnetic biological detection technology has received attention from an increasing number of researchers due to the characteristics of high sensitivity and high detection speed [5]. The basic principle of electromagnetic detection technology is electromagnetic biological effect. Specifically, organisms can produce biological tissue effects when they are affected by the external electromagnetic field. People can get relevant information based on the relevant reactions of organisms. Since the inextricable connection between biological electromagnetic effects and life systems, it can be used t explore the positive effects of electromagnetic fields to promote the development of humanit

More important, the biological action mechanism and biological effects of electromagneti fields are closely related to the excitation frequency, intensity, and duration of electromagneti fields [6]. Different electromagnetic fields have different biological targets, scope of actio action time, and energy and information exchange methods [7]. Therefore, electromagnetic de tection has a wide range of applications. For example, electromagnetic biosensors have playe an important role in increasing the germination rate of seeds [8], improving the adaptability o organisms to the environment [9], controlling environmental pollution [10], etc. At the sam time, bioelectromagnetic detection is also applied in body health monitoring, especially th emergence of nuclear magnetic resonance technology, which has pushed bioelectromagneti detection to a new peak. In addition, treatment methods based on electromagnetic biologica effects have emerged one after another, such as magnetized water and magnetic acupunctur The study of bioelectromagnetic detection is of great significance to promote the level of socia medical care and improve the quality of human life.

This article mainly introduces the biomedical applications of electromagnetic detec tion in recent years. Specifically, this review first introduces the electromagnetic biologica effects and their mechanism of action. In fact, the excitation frequency and intensity ar the main parameters affecting electromagnetic biological effects. Taking the excitatio frequency of the electromagnetic field as a clue, the application of bioelectromagneti detection, including health monitoring and disease treatment, is discussed. Furthermore electromagnetic detection in combination with ML technology has been used in clinica diagnosis since the ML can extract the deep features of electromagnetic signals to obtai deep biological information. The relevant research involving the application of ML tech nology to electromagnetic medical images is summarized. Finally, the author presents a conclusion on the reviewed biosensing techniques, and provides an opinion on the futur development trend of electromagnetic detection.

2. Electromagnetic Biological Effect and Mechanism

When the electromagnetic field acts on an organism, various substances in the or ganism will react to the alternating electric and magnetic fields in the way of induction which in turn results in the biological effect of the electromagnetic field. The biological effects of electromagnetic fields can be divided into thermal effects, non-thermal effects and cumulative effects [11]. The biological thermal effect of electromagnetic wave refers to the biological effect caused by the Joule heat converted from the electromagnetic wave penetrating the biological system [12]. Compared with thermal effects, the non-thermal ef fects are changed in tissues or systems, that is, there is no direct relationship with heat after the organism absorbs electromagnetic radiation energy [13]. In addition, if an organism damaged by thermal and non-thermal effects is not repaired in time, the degree of damage will accumulate when it is exposed to electromagnetic radiation again. The above process is called the cumulative effect.

A series of complex bioelectrical activities are regarded as an important part of the whole organism. The shape, structure, and function of different biological levels will change when a certain intensity of electromagnetic waves penetrate the biological system In fact, the biological action mechanism is closely related to the frequency, intensity, and duration of electromagnetic fields [14]. The scope of action of biological objects, action time, energy and information exchange methods, and exchange values are all different due to the different characteristics of electromagnetic waves, which brings different applications Table 1 summarizes different electromagnetic effects and their applications. In order to analyze and understand the electromagnetic biological effect, it is necessary to summarize and sort out the hypothesis of the electromagnetic biological effect mechanism.

Table 1. Electromagnetic biological effects and related applications.

Electromagnetic Biological Effect	Application	References
Thermal effects	Seed sterilization and inactivation	[15]
	Food storage	[16]
	High frequency electric knife	[17]
	Transcranial magnetic stimulation	[18]
	NMR	[19]
Non-thermal effects	Cardiac pacing	[20]
	Cardiac defibrillation	[21]
	Tumor treatment	[22]
Cumulative effects	Transcranial magnetic stimulation	[23]
	Fracture healing	[24]

2.1. Biological Thermal Effects

Thermal effect refers to the effect produced by electromagnetic radiation heating biological tissues or systems. The strength of the thermal effect is monotonously and positively related to the amount of electromagnetic power absorbed by the biological system. There are three ways to generate the thermal effect, namely, friction heat generation, dielectric loss heat generation, and conduction current heat generation [25].

Friction heat generation is caused by the redistribution of electric charges of water molecules, amino acids, DNA, and other biological molecules under the action of alternating electromagnetic fields [26]. It leads to changes in the structure and function of biomolecules. If the frequency of the alternating electromagnetic field is high, it will have two effects. On the one hand, various polar molecules in the biological system will undergo rapid and periodic changes. In the process of change, these polar molecules collide and rub against the surrounding molecules violently, which generates a large amount of heat energy. On the other hand, various ions in the biological system undergo periodic motion near their equilibrium position due to the action of alternating electromagnetic fields. During the movement, these ions collide with other molecules to generate heat energy.

There are positive and negative charges in the molecule. Under the action of the electric field, these positive and negative charges are subjected to electric field forces in different directions, resulting in corresponding displacements [27,28]. The coupling dipole moment and polarization direction of polar molecules changes due to charge displacement, which may lead to changes in the molecular structure. In fact, it takes time to change the polarization of the dielectric. More specifically, if the electric field changes faster than the polarization direction, the polarization of the dielectric will not keep up with the electric field change, and the electric field energy will be lost.

The last electromagnetic heating effect is conduction current heat generation. There are charged particles that can migrate freely in organisms, which makes the organisms generate an electric current under the action of an electric field; the conduction current passing through the tissue can generate heat. The generation of thermal effects generally requires a higher intensity electromagnetic radiation. At present, the research on the thermal effect mechanism is relatively clear.

2.2. Biological Non-Thermal Effects

There are weak electromagnetic fields in the organs and tissues of organisms. These electromagnetic fields are stable and orderly. The non-thermal effect of electromagnetic field will interfere with the inherent weak electromagnetic field inside of the organism. Once the inherent electromagnetic field in a balanced state is disturbed by the external electromagnetic field, the tissues and organs in the organism are damaged. Compared with the electromagnetic thermal effect, the non-thermal effect has four main characteristics. The first one is coherence, that is, the biological effect can only be produced when the electromagnetic wave parameters and the relevant parameters of the biological system meet a certain deterministic relationship. The second feature is the window characteristics, which mainly consists of frequency windows and intensity windows. The former refers to the effect that the target in the biological system only produces certain discrete electromagnetic waves with a very narrow frequency range. The latter refers to the biological effect that the target in the biological system only produces on certain discrete electromagnetic waves with extremely narrow field strength. Another feature is synergy. The weak electromagnetic field and the metabolism of the biological system can stimulate extremely strong biological effects. The last feature is non-linearity. The intensity of the effect will not increase significantly when the intensity of the electromagnetic wave is greater than a certain threshold. Non-thermal effects can occur under extremely low frequency electromagnetic fields and low-intensity radio frequency radiation. In recent years, the non-thermal effects of electromagnetic fields on biological systems have become a research hotspot due to the extensive application of power equipment and radio frequency radiation.

2.2.1. Coherent Oscillation Theory

Regarding the mechanism of the non-thermal effects of electromagnetic fields, Frohlich first proposed the coherent oscillation theory [29]. The theory believes that biological molecules with dipole characteristics have a fixed vibration frequency due to the interaction of the Coulomb gravitational force. Therefore, the polarized wave is formed, and the coherent effect is produced. When the external electromagnetic field energy is introduced into this fixed vibration, it will be stored in the oscillation mode in a purely mechanical manner. The biomolecules are strongly excited and stay away from the thermal equilibrium state. If the external electromagnetic field close to the coherent oscillation frequency of the biological system acts on the biological system, it will produce two kinds of "frequency window" effects. These two kinds of effects are destructive coherence and constructive coherence, which leads to obvious biological reactions.

2.2.2. Ion Transmembrane Cyclotron Resonance Theory

The theory of ion transmembrane cyclotron resonance is another theory referring the non-thermal effects. The theory considers that charged ions in a static magnetic field move in a circular motion under the action of the Lorenz force [30,31]. In the electromagnetic radiation, the alternating magnetic field parallel to the static magnetic field will increase the angular velocity and orbit radius of the charged particles, which in turn produce cyclotron resonance. When the cyclotron resonance frequency is consistent with the ELF frequency of the intracellular Ca^{2+}, it will induce the receptors on the cell surface and the ions in the transmembrane channel to move circularly or spirally. This phenomenon will affect many enzymes regulated by calmodulin and induce various physiological and biochemical changes. At the same time, the electromagnetic field changes the dipole moment of the ion channel, and the cyclotron motion of the ion in the channel will eventually interfere with the process of ion permeability.

2.2.3. Free Radical Mechanism Theory

The currently commonly recognized mechanism is the free radical pair mechanism [32]. According to the quantum theory, the chemical bond formed between two reactive free radicals is requested to be in a single state. Since electrons have a magnetic moment, the local magnetic field generated by other electrons or atomic nuclei in the molecule may reverse the spin direction of an electron. The radical pair changes from a single state to a triplet state. More important, the external magnetic field will maintain the magnetic moment of the electrons and reduce the possibility of spin reversal, thereby affecting the reaction rate. Although the experiment shows that the reaction exists in biological systems, its biological significance is not yet fully defined [33].

2.3. Cumulative Effects

It takes a certain time for the magnetic field to change the metabolism, function, and structure of the organism. This phenomenon is called hysteresis. The main reason for the hysteresis is the existence of various dipoles, ion genes, and inorganic ions in the biological macromolecule [34]. They are restricted by steric hindrance and potential energy. In addition, only a few stable equilibrium positions are available for the particles in space. They can jump from one equilibrium position to another under the effect of external electric fields. There is also an electric dipole moment between each other because ionic groups and complex ions exist in the form of different chemical bonds. When the electric dipole moment rotates towards the outward electric field, a certain potential barrier must be overcome, so there must be a time course. The electric dipole moment must overcome a certain barrier to rotate towards the external electric field, which must take a certain amount of time. From the energy point of view, a higher the product between the field strength and the action time increases the biological effect, which reflects the cumulative process.

3. Application of Electromagnetic Detection Technology at Different Frequencies

The utilization of electromagnetic biological effects involves many aspects, including disease treatment [35], food quality inspection [36], food sterilization [37], plant cultivation [38], and so on. Since the main parameters affecting the electromagnetic biological effects are frequency and intensity, electromagnetic waves at different frequencies produce different biological effects. According to the interaction mechanisms between biological tissues and electromagnetic fields at different frequencies, the radiated electromagnetic fields are usually divided into four frequency bands: electrostatic field (0 Hz), extremely low frequency (ELF, 0–300 Hz), intermediate frequency (300 Hz–10 MHz), and radio frequency (RF, 10 MHz–300 GHz). The frequency band of the electromagnetic field is shown in Figure 1. Therefore, this review discusses the related application of electromagnetic biological detection by using their working frequency band as a clue.

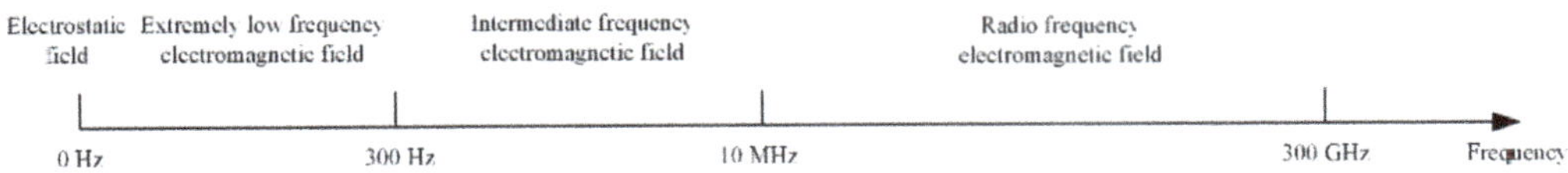

Figure 1. The frequency band of the electromagnetic field.

3.1. High-Voltage Electrostatic Field (0 Hz)

In recent years, high-voltage electrostatic fields have been widely applied in agricultural production and food processing. This technology has an effect on water molecules, enzymes, and other substances inside crop seeds to optimize their original properties and achieve good conditions for seed growth. When the seeds are treated with high-voltage electrostatic fields, ion mist, O_3, and nitrogen oxides will be produced. Nitrogen oxides on the surface of the seed compound with hydration to form nitric acid and nitrous acid, which can stimulate embryos in the dormant state of the seed. Moreover, the O_3 is a strong bactericide, which can greatly reduce the incidence of plant diseases. Therefore, the application of electrostatic field helps to obtain high-yield, high-quality, and non-toxic agricultural products [39]. As for the aspect of food processing, the high-voltage electrostatic field technology can not only be used for the fresh-keeping of fruits and vegetables, but also in food sterilization and thawing to meet human dietary needs.

3.1.1. Seed Germination and Growth

The earliest and most extensive field of electrostatic biological effects research is the use of high-voltage electrostatic fields to treat plant seeds. In 1962, Krueger et al. first discovered that the air environment ionized by positive or negative ions can significantly boost the germination rate of oat seeds [40]. In addition, the growth rate and dry weight of oat seeds after germination are also significantly improved. The experiment equipment i shown in Figure 2 [41]. Sidaway et al. placed lettuce seeds between energized aluminur electrodes for 24 h [42]. The results showed that both positive and negative electrostati fields from 180 V/cm to 360 V/cm can change the germination rate of lettuce seed Figure 3 is the result of high-voltage electrostatic field treatment of safflower seeds, whic further proves the effectiveness of the technology [43]. More important, they found that th positive electric field has an inhibitory effect on plant growth, while the negative electri field has a promoting effect. Huang R. et al. treated cucumber seeds with swelling treatmer and electric field treatment, respectively, and then carried out related experiments abou germination and membrane permeability [44]. The results showed that these two treatmen methods can improve the germination rate of cucumber seeds. In subsequent studie it was further discovered that the combination of a high-voltage electrostatic field an magnetic field technology has a more significant effect on increasing the germination rat of plants. Based on this technology, Li et al. applied a high-voltage electrostatic field wit an intensity of 2.25–2.5 kv/cm to tomato seeds, while applying an alternating magneti field over 8 h [45]. It was found that the germination rate of tomato seeds was increase by 1.1 to 2.8 times compared to the control group. Shao applied Helmholtz coils with th intensity of 2 A to induce the magnetic field. The magnetic field with discharge plasma ca improve the germination rate and activity of old spinach seeds. In fact, the germinatio rate and activity of seeds depend on the flux of the excitation magnetic field [46]. Xu et a applied corona discharge produced by a multi-needle-plate HVEF to treat naked oat seed According to the Fourier Transform infrared spectroscopy (FTIR) of the seed coat, high voltage electrostatic fields treatment can form a new absorption peak at 1740 cm^{-1}, which i closely related to the hydrophilicity of the seed; that is to say, the high-voltage electrostati fields treatment can improve the hydrophilicity of seeds [47].

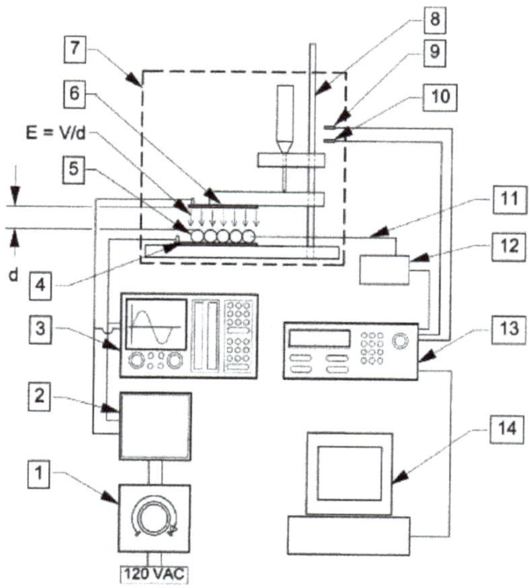

Figure 2. Schematic diagram of experimental set-up for high voltage field treatment of cranberry. Reprinted with permission from ref. [41]. Copyright 2008, Elsevier.

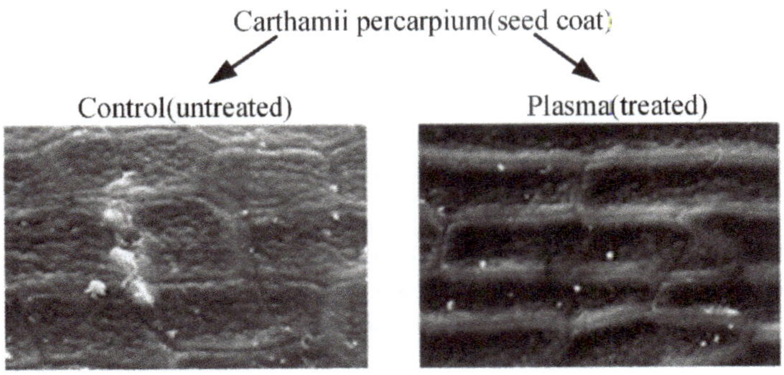

Figure 3. Seed germination processed by a high-voltage electrostatic field. Reprinted with permission from ref. [43]. Copyright 2006 Elsevier.

A high-voltage electrostatic field not only improves the germination rate of seeds, but also improves the physiological indicators of seedlings. Kiatgamjorn et al. found that the germination rate of seeds irradiated with electric fields of 10 kv/m and 25 kv/m are improved, and the seedling height and root length of seedling are significantly higher than seedlings of the control group [48]. In addition, the effect of high-voltage electrostatic field on plant seeds is also manifested in improving the germination index, vigor index, and the activity of various enzymes. Isobe et al. used hydrogen spectrum technology to find that high-voltage electrostatic field treatment can enhance the combined water content of morning glory seeds [49]. In fact, different plants have different sensitivities to a high-voltage electrostatic field environment, and the unsuitable electric field environment may have a negative effect on seed germination. Although Isobe et al. applied a high-voltage electrostatic field of suitable strength to increase the germination rate of morning glory seeds, the germination rate of morning glory seeds is significantly reduced when the electric field intensity is 500 kv/m. A suitable high-voltage electrostatic field can improve seed vigor and make seedlings grow vigorously, thereby improving the ability of seed growth to absorb nutrients and transform and utilize energy substances.

3.1.2. Food Thawing

As shown in Figure 4, the high-voltage electrostatic field thawing technology is a new type of non-thermal processing method. In the small area around the needle electrode, the electric field is used to accelerate the generation of ions. The generated momentum is transferred from the air ions to the neutral air molecules, and corona wind drives a large amount of fluid to the surface of the material, which in turn leads to an increase in the heat transfer coefficient [50,51]. Foods placed between the electrodes are processed by the applied electromagnetic field to achieve thaw. Therefore, the high-voltage electrostatic field thawing has the characteristics of high efficiency and energy saving. Compared with air thawing, high-voltage electrostatic field thawing can shorten the thawing time by half without affecting the thawing quality. Hsieh et al. used the electrostatic field to thaw tilapia meat [52]. They found that the high-voltage electrostatic field can effectively reduce the cooking loss of chicken by 1% to 3% and increase the water holding capacity by 1% to 2%. He et al. discovered the current under negative voltage is greater than the current under positive voltage for the same applied voltage. The corona wind speed changes within a certain range under a given applied voltage and electrode distance, and it is more stable under negative voltage than under positive voltage. Moreover, the current increases with the increase of the applied voltage, and since the corona wind speed increases with the square root of the current, the thawing rate increases under a high voltage and large current [53]. The rapid thawing characteristics of the high-voltage electrostatic field have also attracted the attention of scholars. Dalvi-Isfahan et al. investigated the voltages and freezing temperatures of thawing meat in a high-voltage electrostatic field [54]. The results demonstrated that a slight increase in the hardness value could be seen with an increased voltage or decreased freezing temperature. However, only changing the intensity of the voltage plays a significant role in the equivalent diameter, hardness, and drip loss. Li et al. used the negative electrons generated by the high-voltage electrostatic field to thaw the samples [55]. Defrosting under the −12 kv high-voltage electrostatic field treatment significantly reduced the water loss and microbial contamination of the fish pieces. In addition, it can enhance AMP-deaminase activity, reduce ACP activity, and delay IMP degradation. Therefore, the method can not only effectively maintain the freshness of the meat, but it also has a faster thawing speed. Rahbari et al. discovered that the fibrils of chicken breast stayed relatively intact without any gaps under 2.25 kv/cm. However, the largest deformation and loss of collagen fibrillation network appear under 3 kv/cm. The full results are shown in Figure 5 under different intensities voltages [56].

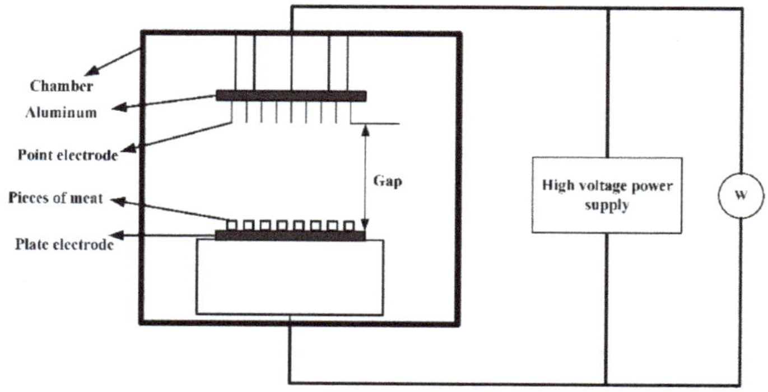

Figure 4. The schematic diagram of high-voltage electrostatic field thawing technology.

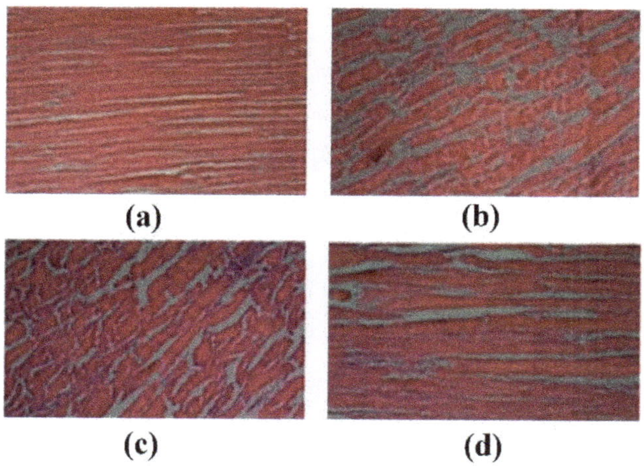

Figure 5. Sections of fresh chicken breast. (**a**) Thawed chicken breast by air method. (**b–d**) Thawed chicken breast by a high-voltage electrostatic field: (**b**) 1.5 kv/cm (**c**) 2.5 kv/cm (**d**) 3 kv/cm. Adapted with permission from ref. [56]. Copyright 2020 Wiley.

As shown in Table 2, the high-voltage electrostatic field has a significant effect on thawing meat products according to the related indicators. Some scholars have discussed the quality of food thawing on the structure and size parameters of the electrode plate. Amiri et al. determined that increasing the number of electrode needles can improve the oxidation degree of beef by fixing the electrode plate spacing [57]. Mousakhani-Ganjeh et al. showed that the oxidation degree of tuna will increase as the distance between electrode plates decreases [58]. Jia et al. found that there is no significant difference between high-voltage electrostatic field thawing and air thawing [59]. Consequently, the oxidation problem caused by the high-voltage electrostatic field can be solved by selecting the appropriate voltage intensity, plate spacing, or the number of electrode plates.

Table 2. Comparison of the thawing meat products under high-voltage electrostatic field.

Material	Sensor Structure Parameters	Thaw Effect	Reference
Beef	4 cm × 4 cm × 2 cm, Space of electrodes: 4 cm, 10 kv	The thawing time decreases as the number of electrodes increases	[39]
Tuna	2 cm × 4 cm × 4 cm, number of electrodes: 16, 5–14 kv	Thawing time is significantly reduced	[38]
Pork	5 cm × 5 cm × 1 cm, Space of electrodes: 5 cm, −10 kv	The pH and tenderness do not present obvious variation from normal air thawing	[40]

3.1.3. Food Preservation

The mechanism of electrostatic field preservation is quite different from that of plant growth. The electrostatic field plays a positive role in the germination of seeds and plants to a certain extent, which strengthens the metabolic process [60]. However, preservation is to delay the aging process of fruits and vegetables by inhibiting the respiratory metabolism of fruits and vegetables. There is still a big controversy in the preservation mechanism. Some researchers believe that the electric field affects the metabolic activity of biological cell membranes by changing the transmembrane potential. Others hold the view that the biological electric field inside the fruits and vegetables affects the electron transporter in the respiratory system.

Zhao et al. used a high-voltage electrostatic field to treat green mature tomatoes for 2 h at 20 °C. The high electric field apparatus contains a power generator, treatment chamber, two stainless steel plates, one high-voltage controller, one voltmeter, and one amperometer. It can output −30–0 kv and 0–30 kv maximum, respectively. The results shown that the respiration rate of vegetables after electric field treatment decreased, which prolonged its storage time [61]. Kao et al. treated fresh-cut broccoli with 50–400 kv/m high-voltage electric field at 4 °C. The results show that the high-voltage electric field can prolong the storage time of fresh-cut broccoli according to the analysis of color and vitamins [62]. Takaki et al. introduced vegetables and fruits with a high-voltage electrostatic field. Compared with the control group, the storage time of the juice after the treatment was prolonged [63]. As for meat products, Wen-Ching Koa et al. used 100 kv/m high-voltage electrostatic treatment on tilapia. The equipment included an electrostatic generator, whose output voltage was 50 kv. This technology can delay the water and salt soluble to prolong the fresh-keeping period of fish by ~2 days [64]. Moreover, the 300 kv/m high-voltage electrostatic can inhibit actomyosin Ca^{2+}-ATPase activity. If a 600 kvrt/m high-voltage electrostatic is applied, a longer extension time can be obtained.

3.2. *Extremely Low Frequency Electromagnetic Field (0–300 Hz)*

The extremely low frequency electromagnetic field (ELF-EMF) is the most common electromagnetic field because the power frequency of most countries in the world is in this frequency band. ELF-EMF has a certain effect on humans, including immune function, cell differentiation, cell proliferation, and transmission of intracellular signal substances. It can achieve cell differentiation, tissue repair, differentiation of stem cells, and healing of injured cells. Specific ELF-EMF can stimulate the production of local bone factors, and significantly improve the proliferation and differentiation indicators of bone density. In addition, ELF-EMF can also inhibit tumor growth and induce cell apoptosis [65,66]. More important, the human nervous system is the most sensitive issue to ELF-EMF. A large number of experiments have shown that the effect of ELF-EMF on the central nervous system is double-sided. On the one hand, the ELF-EMF may affect the normal brain development. It has an important impact on the growth and differentiation of neurons and the direction of neurites. On the other hand, electromagnetic fields may cause oxidative stress in the brain tissue, which leads to the activation of astrocytes. Based on the above research, this review focuses on the application of extremely low frequency electromagnetic sensors in cancer treatment, fracture healing, peripheral nerve repair, and transcranial magnetic stimulation.

3.2.1. Cancer Treatment

ELF-EMF can induce apoptosis of cancer cells to some extent, thus providing a new way for the treatment of cancer. Studies have shown that ELF-EMF have significant therapeutic effects on a variety of tumors. They can induce growth inhibition and tumor killing effects on lung cancer, liver cancer, breast cancer, prostate cancer, and melanoma cells [67–69]. The main anti-tumor mechanisms of electromagnetic fields can be roughly divided into the following four categories:

1. Induce the level of ROS in tumor cells to increase, which in turn damages DNA, proteins, and membrane lipids;
2. Produce selective cytotoxicity to tumor cells and achieve an anti-tumor effect by improving cellular immunity;
3. Directly damage the DNA chain to cause chromosome aberrations and inhibit tumor growth;
4. Promote tumor cell apoptosis and cell cycle arrest.

The rotating electromagnetic sensor has a significant effect on tumor suppression. In 2011, Wang et al. found that the magnetic field with an intensity of 0.4 T and excitation frequency of 7.5 Hz for 4 days can inhibit the proliferation of a variety of human tumor cells including gastric cancer BGC-823, gastric cancer MKN-28 cells, and lung cancer A549 [70]. However, this technology does not inhibit poorly differentiated gastric cancer MKN-45 cells and lung adenocarcinoma SPC-A1 cells. It cannot inhibit poorly differentiated gastric cancer MKN-45 cells and lung adenocarcinoma SPC-A1 cells. Ren et al. found that the rotating magnetic field with an intensity of 0.4 T and excitation frequency of 7.5 Hz for 6 days can induce the senescence of lung cancer cell A549 and inhibits its iron metabolism [71]. If the magnetic field intensity and frequency are changed, for example, the rotating magnetic field with an intensity of 0.4 T and excitation frequency of 7.5 Hz for 15 min does not have a significant effect on the proliferation of HepG2 liver cancer cells. The research shows that the intensity of the magnetic field, the processing time and the cell type are all key factors for the effect of magnetic field on cells.

Zhang et al. designed a gyromagnetic sensor composed of two groups of 30 neodymium iron boron permanent magnets which are anti-parallelly stacked. The structure of sensor presented in Figure 6a [72]. The magnet array rotates on the vertical axis at a frequency of 8–10 Hz. The MF intensity at the top of the processing table is 0.6 T, and the MF intensity at 5 cm above the surface is 0.4 T. Moreover, the horizontal spatial change in the animal cage region is 0.6–0.32 T. The rotation introduces an approximately sinusoidal 0.21 T average fluctuation in the MF intensity with a frequency of 8–10 Hz. As shown in Figure 6b, Du et al. investigated the gyromagnetic sensor with a cylindrical permanent magnet rotating around the long axis [73]. Some scholars studied the gyromagnetic sensor composed of NdFeB Halbach array magnets. As shown in Figure 6c,d, a uniform 1 T magnetic field is generated radially in the center part of the device, and a magnet is installed on the motor to control its rotation [74]. Under a rotating magnetic field, the internalized MPs can effectively destroy the cell membrane through mechanical force and promote programmed cell death in vitro and in vivo. Magnetic field therapy successfully reduced the size of brain tumors and prolonged survival rates. The development of the above-mentioned sensors can significantly improve the uniformity of the magnetic field and accelerate the induction of apoptosis of some cancer cells to a certain extent.

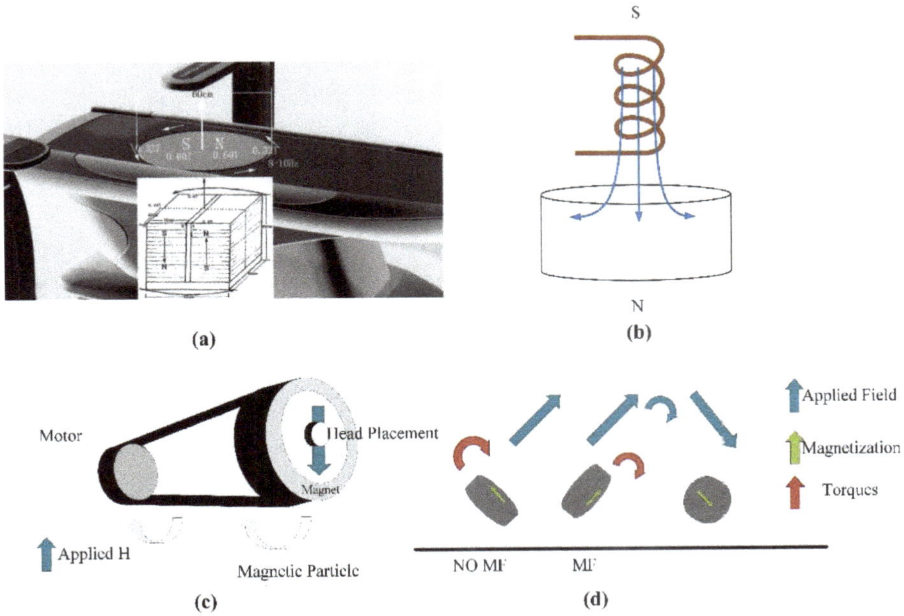

Figure 6. The structure of the gyromagnetic sensor. (**a**) Two anti-parallel stacks of 30 neodymium-iron-boron permanent magnets. Reprinted with permission from ref. [72]. Copyright 2005 Wiley-Liss. (**b**) Cylindrical permanent magnet rotating around the long axis. (**c**) NdFeB Halbach array magnet. (**d**) The changes of the magnetic field corresponding to (**c**)

Sisken et al. used pulsed electromagnetic fields with an intensity of 0.3 mT and a frequency of 2 Hz to treat rat sciatic nerve defects, which proves that electromagnetic fields can stimulate the regeneration of the nervous system [75]. Experimental results show that the application of pulsed electromagnetic fields can accelerate the formation of myelin and connective tissue [76]. Zhong et al. measured the proliferation and differentiation of bone marrow mesenchymal stem cells under condition of the electromagnetic field with an intensity of 0.5 mT and frequency of 50 Hz [77]. The results confirmed that prolonging the stimulation time can promote the cell proliferation and induce the cell differentiation. After transplantation into animals, the original cells of bone marrow mesenchymal stem cells stimulated by the magnetic field performed obvious biological characteristics. Diniz et al. used a Helmholtz copper coil to analyze the effect of pulsed magnetic fields on the proliferation of osteoblasts, which also confirmed the effectiveness of electromagnetic stimulation. The diameter and number of turns are set to 12 cm and 250, respectively. The excitation frequency of the coil is set to 15 Hz [78]. Liu et al. designed a set of ELF-EMF biosensors consisting of a driving unit and a pair of Helmholtz coils. The diameter of the coil is 14 cm, and the distance between the coils is 13 cm. The induced peak voltage and the change intensity of the magnetic field are ±0.7 mV and 0.4 T/s, respectively [79]. They found that the pulsed magnetic field generated by the electromagnetic sensor can protect the synthesis of chicken sternum cartilage proteoglycan in the embryo.

3.2.2. Transcranial Magnetic Stimulation (TMS)

TMS is based on the principle of electromagnetic induction, which applies a rapidly changing current to the coil to generate a strong and short-lived magnetic field, as shown in Figure 7 [80]. When the voltage of the induced electric field reaches a certain strength, the resting membrane potential of the cell will change. Thereby, it has an excitatory or inhibitory effect on the nervous tissue to affect the physiological and biochemical activities of the brain, and finally achieves the therapeutic effect on neurological and mental diseases.

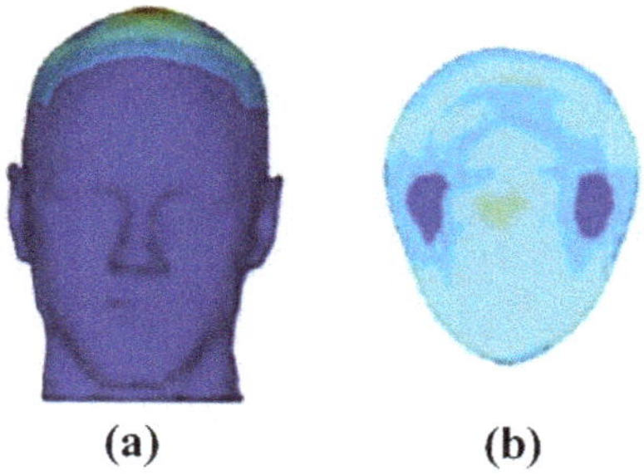

Figure 7. Magnetic field distribution stimulated by TMS. (**a**) Front view of the induced electric field distribution on the scalp surface. (**b**) Sectional view of the x-y plane.

In the application of cranial nerve stimulation, the magnetic field generates an induced electric field in the brain tissue to form an intracranial current, thereby changing the stimulation location and related neuronal activity. In 1985, Anthony Baker et al. were the first to study TMS. Nowadays, TMS has been widely used in clinical experiments and research in the fields of neurology, neuroscience, psychiatry, rehabilitation, psychocognitive science, and sports medicine [81]. It has a good therapeutic effect in depression, obsessive-compulsive disorder, schizophrenia, and peripheral nerve rehabilitation [82].

Jiang et al. used TMS to treat 40 patients with chronic primary insomnia. The results show that TMS treatment significantly improved the sleep cycle of stage III sleep and rapid eye movement sleep [83]. Levitt et al. used 10 Hz TMS to treat patients with refractory depression [84]. Magnetic resonance spectroscopy is used to assess the changes in GABA levels of the left dorsolateral prefrontal cortex (DLPFC). The results indicate that TMS treatment is associated with elevated GABA levels, and subjects who receive GABA agonist treatment have less response to TMS. Leblhuber et al. studied the changes in the utilization of neurotransmitter precursor amino acids in elderly patients with depression after stimulation of the prefrontal cortex, and found that TMS has a significant effect on the treatment of depression [85].

The performance of the TMS system is mainly determined by the focus, stimulation depth, and intensity of the induced electric field. In fact, the structure of the TMS coil is directly related to the electric field distribution in the brain tissue. Among them, the depth and focus of the induced electric field distribution are the two most critical indicators for the optimal design of the TMS coil [86]. In order to locate the optimal stimulation point of the sensor to the intracranial cerebral cortex, the structure of the sensor needs to be precisely designed. It is helpful to make continuous breakthroughs in the performance of TMS by simulating the performance of the coil in terms of stimulus intensity, stimulus depth and focus.

(1) Circular coil

The structure of the circular coil is simple, and the induced electric field on the surface of the brain is distributed in a "circular shape." The stimulation performance of the circular coil is poor, especially in focus behavior. In order to improve the focus, one side of the coil can be folded up along the center of the circle. The novel structure of coil is called the deformed circular coil. The simulation results show that the magnetic field induced by the deformed circular coil has a large attenuation. The change in the magnetic field leads to a change in the electric field distribution, which in turn results in an improved focusing. Figure 8a shows the structure of a circular coil and a deformed circular coil.

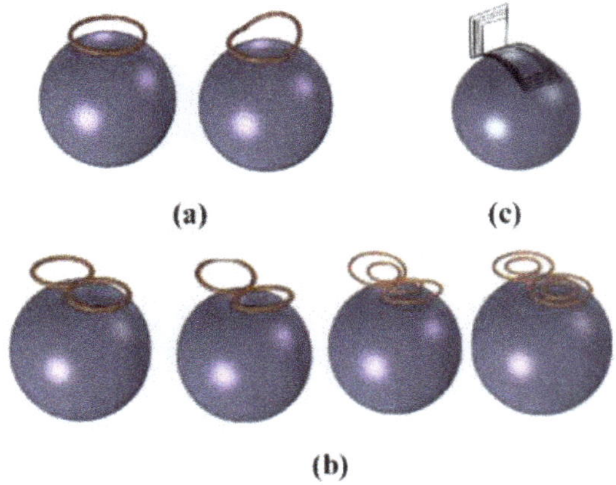

Figure 8. The structure of the TMS coil. (**a**) Circular coil. (**b**) Figure-of-eight coil. (**c**) H-shaped coil.

(2) Figure-of-Eight coil

Figure-of-eight coils are widely applied in commercial applications. Two coils with opposite currents cause the peak of the induced electric field to appear at the tangent position of the coil [87,88]. Figure 8b shows the figure-of-eight coil. Compared with the circular coil, the focus of the figure-of-eight coil is greatly improved. However, it can only be applied to the stimulation of the superficial areas of the brain. In order to further improve the performance of the figure-of-eight coil, a butterfly coil and a quadruple butterfly coil are proposed. The structure of the butterfly coil and quadruple butterfly coil are improved based on the figure-of-eight coil. By enhancing the magnetic field at the tangent of the coil, the focusing performance is improved and the target point has greater stimulation intensity. In addition, the quadruple butterfly coil can be used in combination with the shielding plate, which can further reduce the focal area of the brain surface.

(3) H-shaped coil

The H-shaped coil does not cause high electric field strength in non-target areas because it has a small electric field attenuation rate and surface charge accumulation. As shown in Figure 8c, it can effectively stimulate the deep neuron area, so it belongs to the deep stimulation coil. In addition, the penetration depth of each area can be flexibly controlled by adjusting the coil input or changing the distance between the coil and the brain [89,90].

(4) Halo coil

A large diameter circular coil placed in the middle of the head is called the Halo coil. The special position of the Halo coil enables it to stimulate deeper tissues in the brain; thus, the Halo coil belongs to the deep stimulation coil category. Figure 9 and Table 3 show the structure and model characteristics of the Halo coil, respectively. Halo coils can also be combined with other coils, such as the Halo-circular assembly coil (HCA coil), Halo coil working with two circular coils (HTC coil), Halo and Fo8 coils (HFA coil), double-cone coil with the Halo coil (HAD coil), and the triple-halo coil (THC coil) [90–92]. Although different coils have different structures and characteristics, they all realize the superposition and cancellation of the magnetic field by changing the direction of the input current, which increases or reduces the intensity of the induced electric field.

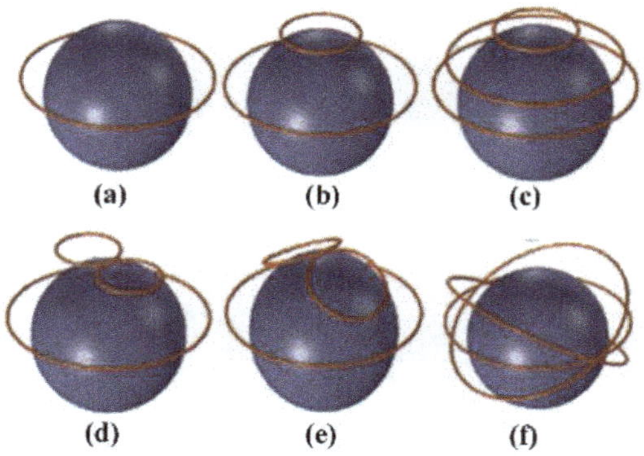

Figure 9. The structure of the Halo coil and related combinations. (**a**) Halo coil. (**b**) HCA coil. (**c**) HTC coil. (**d**) HFA coil. (**e**) HDA coil. (**f**) THC coil.

Table 3. Characteristics of the Halo coil model.

Coil Model	Advantages	Disadvantages
Halo	Increase the electromagnetic penetration depth	The required the current is relatively large
HCA	Improve the penetration depth	Poor deep focality
HTC	Better deep focality	Compared with HCA, the strength of the induced electric field is reduced
HFA	The electric field strength and penetration depth increase on the side	The focality is poor and the attenuation rate is increased
HAD	Compared with HFA, the strength and penetration are larger	The focality at the gray and white areas is poor
THC	High flexibility	The superficial electric field strength is high

Moreover, there are linear coil arrays, cap-formed coil arrays, multi-layer coil arrays, etc. [93,94]. The structures of the coil arrays are shown in Figure 10. The linear coil array has a small number of input control units, and the stimulation mode is flexible and changeable by changing the size and direction of the input current. The cap-formed coil array has a greater stimulation depth and weakens the impact on non-target areas due to the increased coupling with the head surface. The multi-layer coil array increases the stimulation intensity to a certain extent. As the number of coils increases, the number of channels of the magnetic field generator should also increase accordingly. In other words, the overall structure and working conditions are more complicated, and hence, the multi-layer coil array has poor applicability.

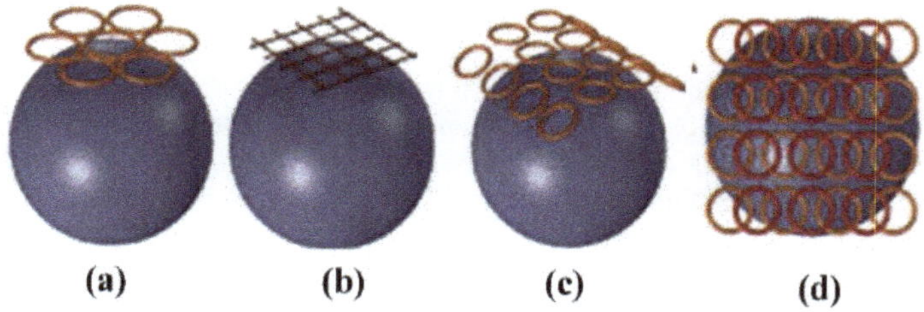

Figure 10. The structure of the TMS coil arrays. (**a**) Circular coil array. (**b**) Linear coil array. (**c**) Cap-formed coil array. (**d**) Multi-layer coil array.

3.3. *Intermediate Frequency Electromagnetic Field (300 Hz–10 MHz)*

Intermediate frequency electromagnetic detection are widely applied in areas such as food quality and spatial location monitoring. In food quality monitoring, the organization, composition, structure and state of meat products, fruits, and vegetables are closely related to their electromagnetic properties [95]. For example, the impedance value has a certain change rule in the process of meat decay. In addition, the impedance value of the electromagnetic sensor is also closely related to its position, so the electromagnetic sensor can detect changes in the relative position in space. At present, this technology is mainly used for the monitoring of human joints, limbs, and eye movements. The applications of electromagnetic sensors in food quality and spatial location monitoring are introduced as follows.

3.3.1. Food Quality Inspection

The decay of meat products produces many complex biochemical and physical and chemical processes, which will affect its organization and structural characteristics [96]. These phenomena also change the electrical parameters. Biological tissue can be considered as a non-magnetic material, and its electrical properties are the result of the joint action of its tissue and structure. Among them, the most important component that affects the dielectric properties of the tissue is the ion. As for the structure of the dipole moment, the main factors are water tissue molecules, proteins, and lipids. The movement of electric charges causes conduction effects, and the polarization of dipoles produces dielectric relaxation. The conductivity of most tissues is low when electromagnetic waves are excited at low frequencies, which mainly depends on the volume fraction of extracellular fluid. As the excitation frequency increases, the conductivity of intracellular and extracellular ions rises sharply due to the dielectric relaxation of water. The whole process can be represented by Figure 11.

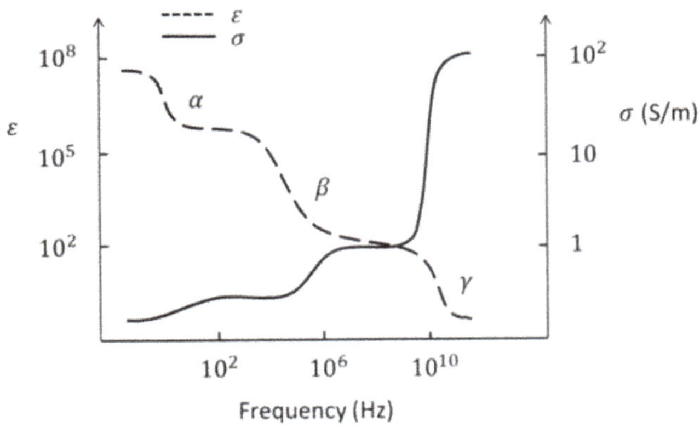

Figure 11. The dispersions of biological samples.

The above-mentioned increase in conductivity is called dispersion. In biological systems, there are mainly three kinds of dispersions (α, β, γ). According to the phenomenon of dispersion, it is divided into three frequency bands. The frequency range of α dispersion is from millihertz to hundreds of hertz, and it is mainly related to the phenomenon of tissue polarization. The frequency range of β dispersion caused by the Maxwell-Wagner effect is from hundreds of hertz to megahertz. It is a method of measuring the integrity of cell membranes during meat aging because of reduced insulation properties. The frequency range of the γ-dispersion is in gigahertz, which is mainly caused by the permanent dipole relaxation of free water in muscle tissue. This characteristic also further proves the reliability of the electrical method for detecting meat products. In fact, electrical detection can be divided into contact measurement methods and non-contact measurement methods. The response speed of the electrode contact measurement method is fast and accurate. However, the contact method may cause unnecessary measurement errors, such as the chemical reaction and food tissue damage. This problem does not exist in non-contact measurements. As a non-contact measurement method, the electromagnetic detection method has a wide range of applications in food quality detection [97].

Swatland et al. reported the changes in rheological properties, electrical properties, and optical properties during the ossification of pig and cattle carcasses [98]. This process can be detected by its low reactance and low capacitance, and it is found that these three physical detection characteristics have similar variation curves. Ibba et al. studied the influence of frequency on the bio-impedance of fruits using electrical impedance spectroscopy. They pointed out that the bio-impedance can be used to measure fruit quality. The results demonstrated the potential of this technique for monitoring the ripening of fruit, obtaining a good correlation of the impedance evolution with low frequency points [99]. Sun et al. combined chemometrics and impedance characteristics to create a multi-mass index for Atlantic salmon. Moreover, they applied a partial least squares method to create a quantitative prediction model of the bioimpedance spectroscopy and total volatile basic nitrogen value. The correlation coefficients of the training set and test set are 0.9447 and 0.9387, respectively, which means that the impedance characteristics measurement method has the advantages of convenience, economy, and high precision. Castro-Giraldez et al. noted that the certain frequencies (500 Hz, 300 kHz) are effective at 24 h after slaughtering to detect the dark firm dry (DFD) meats during the post-mortem period. In fact, there may be systematic errors in the measurement [100]. Kim positioned the monopolar injection needle to the dermis, subcutaneous tissue, or muscle layer of pork tissue, and then measured electrical impedance in the frequency range of 10 Hz to 10 kHz based on an impedance analyzer [101]. This method has also been proven to be used for pork quality monitoring. Tang et al. introduced a non-contact induction measurement system for bioimpedance spectroscopy [102–104]. The electrical conductivity of the food is measured by an excitation-retraction electromagnetic sensor in the frequency sweeping mode, and the freshness of the food is detected according to the change of the electrical conductivity signal. Apart from that, Tang et al. used the fast finite element method to establish the cell model, as shown in Figure 12. The solution of the numerical model can obtain the vortex field distribution inside the cell and further proves the accuracy of the measurement results.

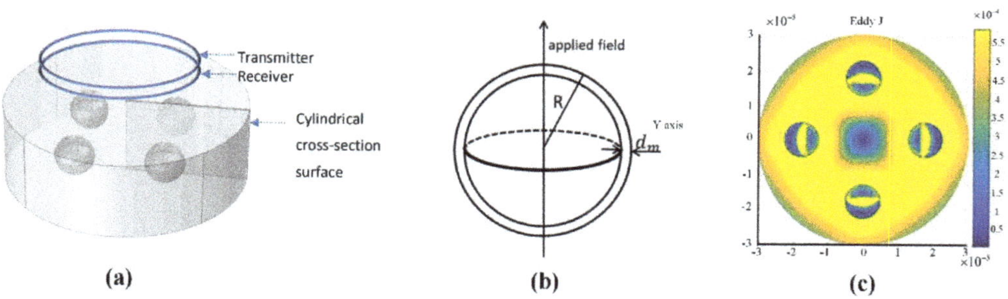

Figure 12. Three-dimension finite element model. (**a**) Measurement set-up. (**b**) Conductivity spectroscopy of fresh potato and defrosted potato. (**c**) Simulation model. Adapted with permission from ref. [102]. Copyright 2019 Elsevier.

3.3.2. Spatial Position Measurement

Electromagnetic sensors can be used for spatial position measurement. When a small receiving sensor moves in space, the electromagnetic positioning system can accurately calculate its position and orientation, thereby providing dynamic and real-time measurement of the position and orientation angle [105,106]. In biomedical research, it is an ideal choice for measuring range of motion and limb rotation due to the advantages of being fast and precise. The electromagnetic tracking system generally consists of a magnetic field transmitter and receiver. A magnetic field generator is placed near the target, and the magnetic field covers a certain range. The receiver detects the strength and phase of the magnetic field and sends the signal to the control unit. The control unit can calculate multiple degrees of freedom for tracking the target. The advantage of the electromagnetic tracking system is that the detection result is not restricted by the line of sight. Nothing can block the tracking of the electromagnetic tracking system except for electrical conductors or permeable magnets. The disadvantage is that it is easily interfered by metal and the working range is small.

Recently, a three-dimensional electromagnetic tracking system was launched, as shown in Figure 13. It can measure the range of motion of the lumbar spine, cervical spine, and shoulders of humans. The "Flock of Birds" electromagnetic tracking device is used for three-dimensional motion analysis, and the patient's arm is extended in three directions [107]. The slope of the regression line between the rotation of the scapula joint and the rotation of the scapula is calculated, which reflects the rhythm of the scapula. According to the measurement result, the movement information of the shoulder joint can be obtained in time.

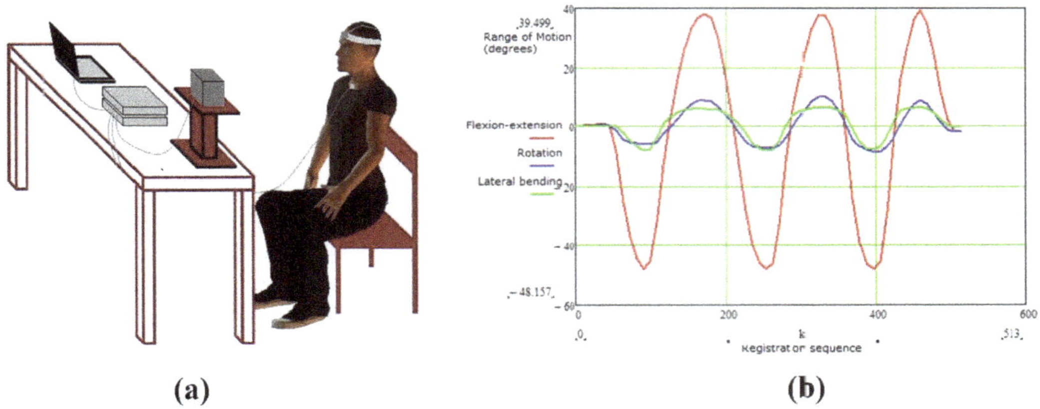

Figure 13. Three-dimensional electromagnetic tracking system. (**a**) Schematic diagram of detection device. (**b**) Movement measurement signal. Adapted with permission from ref. [107]. Copyright 2018 Elsevier.

In accordance with the idea, Xie et al. proposed a new method of biomechanical motion monitoring based on electromagnetic sensing technology [108]. The proposed method has the advantages of non-invasiveness, easy implementation, low cost, and high efficiency. A theoretical model was established to model the sensor response, and the wearable sensor system and corresponding measurement signal shown in Figure 14 was designed to monitor various biomechanical movements, including blinking frequency, finger/wrist bending level, and frequency. The whole system is measured by the mutual impedance of the coil. According to the change frequency of the signal per unit time, it can reflect the body's movement state in time. Therefore, this technology can be used to detect early signs of these abnormal biomechanical behaviors, so that relevant treatment can be carried out in time. Robison invented an electromagnetic method of eye movement recording, which has the characteristic of low noise; however, the method is limited by blinking and fluctuations in pupil size [109]. Remmel designed a cheaper version. The system uses three alternating magnetic fields (48, 60, and 80 kHz) in the X, Y, and Z directions to record eye movement information [110]. The subject wears a contact lens with a small loop of wire, and his eyes are located close to the center of the test field. The three alternating magnetic fields introduce three voltages into the coil through Faraday's law of induction. As the eye rotation, more magnetic field passes through the coil, which generates induces larger voltage. Three analog (direct current) voltages of the vector components in the X, Y, and Z directions are obtained by amplifying and demodulating the coil signal. According to the measurement signal, the angle information of the eye can be inverted.

Later, Remmel extended this method to arm recording [111]. The two small coils fixed a the elbow and two coils fixed at the wrist are applied to measure the movement of the arm For each coil in the system, the voltage is demodulated to give three DC voltages to obtai the vector direction of the coil axis. At the same time, the inhomogeneity of the magneti field is corrected. This method can describe five angles of arm movement, and the noise i only 2 arcsec rms.

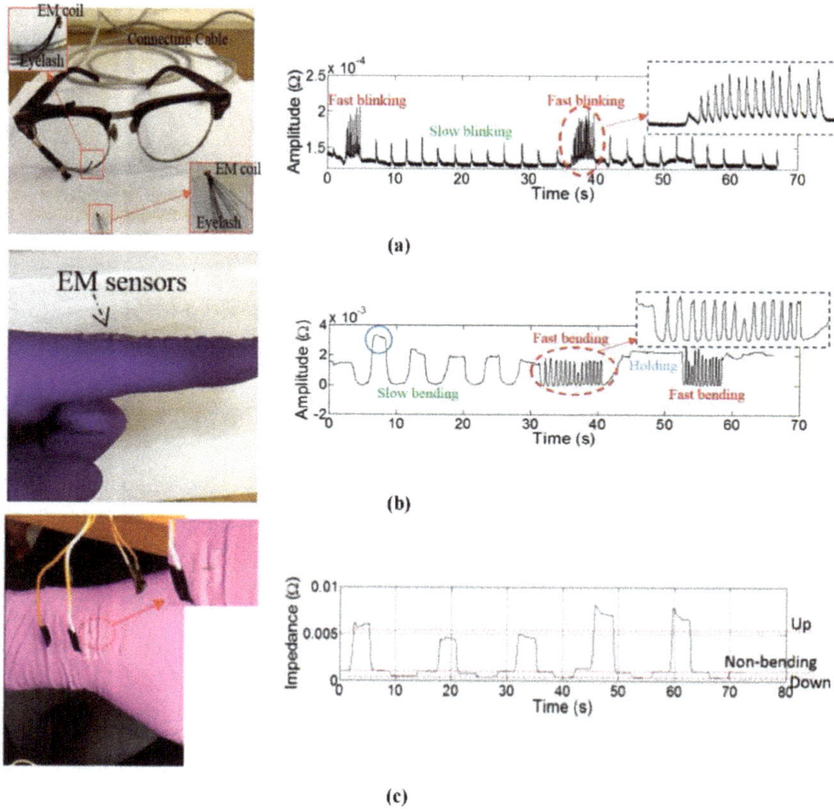

Figure 14. The biomechanical motion monitoring based on electromagnetic biosensors. (**a**) Wearable EM sensor. (**b**) Finger bending. (**c**) Wrist bending. Adapted with permission from ref. [108]. Copyright 2020 IEEE.

3.4. *Radio Frequency Electromagnetic Field (10 MHz–300 GHz)*

Radio frequency electromagnetic fields (RF-EMFs) generated by communication base stations, mobile phones, wireless phones, and wireless routers have become an indispens able part of modern life. The non-thermal effects of radio frequency electromagnetic fields have a serious impact on daily lives [112]. Although the non-thermal effects of radio frequency electromagnetic fields have a serious impact on people's daily life, electromag netic fields can react with biological tissues for disease detection. Therefore, this review will introduce some applications and developments of high-frequency electromagnetic detection in disease detection.

3.4.1. Disease Detection Based on MRI

Nuclear magnetic resonance (NMR) technology is mainly based on the interaction between magnetic nuclei and magnetic fields. The nuclear magnetic moment of the low energy state absorbs the energy provided by the alternating field strength, and then continuously transforms to the high-energy state, thereby generating nuclear magnetic resonance signals [113]. This transition phenomenon caused by resonance absorption is called the nuclear magnetic resonance phenomenon.

MRI is based on the nuclear magnetic resonance signal emitted by the hydrogen nuclei in the human body. The strength of this signal depends on the position of the hydrogen nucleus in the molecular structure and the surrounding environment, so it can reflect a large amount of media information. The strength of this signal depends on the position of the hydrogen nucleus in the molecular structure and the surrounding environment, so it can reflect a large amount of media information. In the late 1980s, the phased array coil concept was introduced into the field of magnetic resonance imaging, which revolutionized the reception of MR signals [114]. The receiving array can use multiple small surface RF coils to obtain a higher signal-to-noise ratio image in a wider range by appropriately combining the signals from a single element. Since the efficiency of parallel imaging and the achievable acceleration factor depend on the number of coils in the receiving array, the number of receiving channels in the MR system has been increasing over time [115]. In fact, most research is based on phased array coil technology to reduce the coil size and increase the number of channels, which can improve the imaging quality [116]. Compared with traditional mechanical coils, wearable technology can maximize the fill factor of the magnetic resonance receiving coil to significantly improve the imaging quality [117].

The Institute of Biomedical Engineering of the University of Zurich in Switzerland has made various attempts to implement wearable technology [118,119]. As shown in Figure 15, the fabricated coil adopts an adjustable mechanical structure to adapt to the size change of the measured part. This kind of coil has a complex structure and large volume, which determines that it is difficult to have a wide range of applications. Subsequently, the team used thin wires to achieve the softness and scalability in a single direction by braiding a wire harness. In in vivo knee imaging, it was demonstrated that a stretchable array of eight receiver coils allows adaptation to different anatomical dimensions and different knee flexion angles.

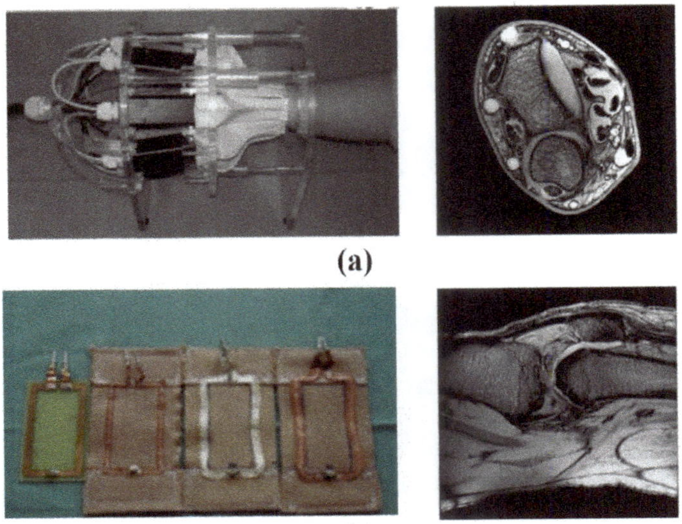

Figure 15. The adjustable mechanical structure for MRI. (**a**) Wrist coil and corresponding detection results. Adapted with permission from ref. [118]. Copyright 2009 Wiley-Liss. (**b**) Knee joint coil, consisting of the reference coil made from circuit and stretchable coils made from 5 mm, 7 mm, and 9 mm copper braids, and detection results. Adapted with permission from ref. [119]. Copyright 2012 Wiley-Liss.

In order to explore the relationship between the number of phased array coils and the ultimate performance, the GE global research team created a 128-channel "flexible" coil for torso imaging. As shown in Figure 16a, this coil can change the shape of the coil within a certain range. However, the huge geometric size and limited deformation space are not suitable for infants and weak patients [117]. The signal-to-noise ratio (SNR) in the proposed system is 1.03 in the center of the elliptical loading model and 1.7 in the outer region on average. Furthermore, the maximum g factor of 4 × 4 acceleration is reduced to 2.0, and the 5 × 5 acceleration is reduced to 3.3. These characteristics significantly reduce the residual aliasing artifacts in human imaging.

As shown in Figure 16b, Jia et al. deployed coils on light and soft materials, but the coil array cannot be stretched. The feature of extremely flexible makes it suitable for inspections in many scenes [120]. The experiments show that the proposed MRI system can achieve a high (SNR) of 336 in vitro, and the phantom images are evenly distributed. As for in vivo experiments, it has a higher SNR of 169 in the region of interest, and the longitudinal coverage exceeds 48.5 cm. However, the bulky system line is still the biggest obstacle affecting its application. As shown in Figure 16c, Corea et al. used inkjet printing and screen-printing techniques to make coils [121]. At the same time, the relationship between the printing coil inductance and the "distributed capacitance" in different printing materials was analyzed. The results shown that the relative dielectric constant increases linearly with the concentration of barium titanate in the ink. In addition, the fabricated coil can be attached to the tested tissue, which is beneficial to improve SNR. The cost of this technology is relatively high, and the coils made by printed materials are less flexible. Therefore, materials that can meet the printing technology and high flexibility are still the bottleneck of the realization of this magnetic resonance coil.

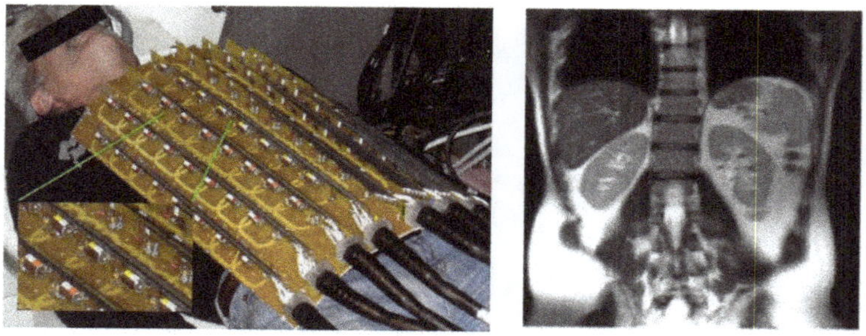

(a)

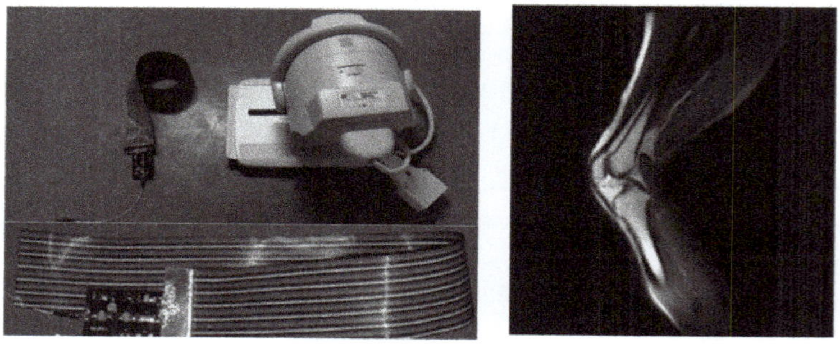

(b)

Figure 16. *Conts.*

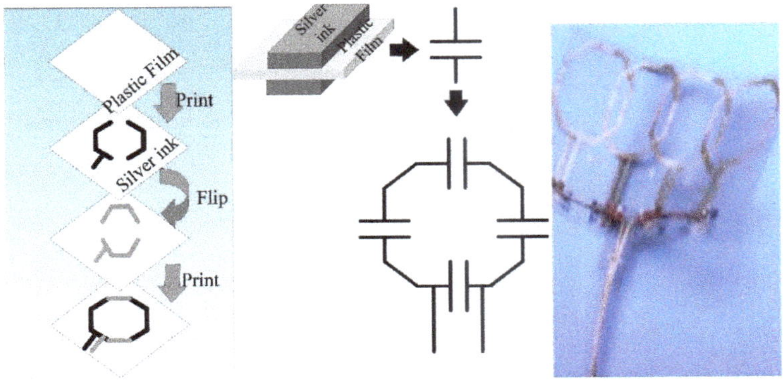

(c) (d) (e)

Figure 16. The wearable MRI sensor. (**a**) The 128-channel torso array coils and corresponding imaging results. Adapted with permission from ref. [117]. Copyright 2008 Wiley-Liss. (**b**) Lightness and softness coils and corresponding imaging results. Adapted with permission from ref. [120]. Copyright 2015 John Wiley & Sons. (**c**) Coil printing process flow. (**d**) Schematic of a printed coil. (**e**) Photograph of a printed coil.

3.4.2. Disease Detection Based on Microwave

The main task of microwave detection is to simulate the scattering and diffraction of the electromagnetic field passing through the medium [122]. This technology measures the electromagnetic field scattering information of the object medium to reconstruct the dielectric properties and geometric characteristics of the object. Since the quality of the microwave image is greatly affected by the dielectric properties of the tested target, the existing microwave imaging methods are mainly used for breast cancer detection. Only a few studies are focused on the detection of strokes.

In breast cancer detection, Wang introduced a fast microwave sensor for breast cancer detection through near-field imaging methods [123]. The method of detecting cancer cells based on microwave imaging is highly correlated with the contrast of dielectric properties between healthy tissue and malignant tissue.

Bassi et al. investigated the confocal microwave imaging (CMI) for breast tumor detection. The results show that the 2D CMI system can detect small tumors (diameter 2 mm), and the 3D CMI system can identify tumors of medium size (diameter greater than 6 mm). Moreover, they applied delayed multiplication and summation methods in CMI to reduce artifacts and noise and enhance the image [124]. Yang et al. first proposed the detection method of the Microwave Image for breast cancer detection based on the principle of ground penetrating radar [125]. It uses the transmitting antenna to emit Ultra-Wideband (UWB) microwaves to irradiate the breast, and then the receiving antenna to receive the scattered signal. According to the scattered signal, the strong scattering area in the breast is recovered to determine the position of the tumor. Rappaport et al. used the time reversal algorithm of Finite-Difference Time-Domain (FDTD) to realize the spatial reconstruction of the dielectric constant distribution inside the breast. They deduced the iterative formula of FDTD to obtain the initial distribution of the electromagnetic field in space [126,127]. The method is an ill-conditioned solution problem, for which huge challenges exist in robustness and computational cost.

As for brain stroke detection, researchers at the University of Queensland in Australia have designed an ultra-wideband microwave system for head imaging [128]. According to the construction of the human brain model, the confocal algorithm is used to realize the detection of stroke. Ireland et al. used the Born iterative method to detect hemorrhagic stroke in brain tissue [129]. Furthermore, a high-contrast brain conductivity distribution map is reconstructed through microwave imaging technology. The internal information can be detected according to the distribution of brain conductivity. The reconstruction results of conductivity can be used to detect strokes.

4. Application of Machine Learning Technology in Electromagnetic Medical Images

Artificial intelligence is a technological science which can be used to simulate and expand the theories and methods of human intelligence. Machine learning (ML) is a method of realizing artificial intelligence, which applies algorithms to analyze data, make decisions, and predict specific events. Deep learning (DL) is a technology that realizes machine learning. It is a data-driven method of automatically learning high-level features hidden in images, which can greatly reduce the interference of subjective factors in feature selection [130]. In addition, the model applies a non-linear layer structure, which can be used to build a more complex model. Shallow neural networks are prone to under-fitting while DL can improve the learning ability by increasing the depth of the network to solve more complex problems.

According to the electromagnetic biological effect, the signal detected by the electromagnetic biosensor can reflect the information of the biological body. However, manual analysis and interpretation are still required. In fact, machine learning can be used to learn the characteristics of biological signals, instead of manual operation. In particular, machine learning algorithms have been widely used in medical image diagnosis. Therefore, this review summarizes the application of machine learning in microwave breast cancer images, MRI prostate cancer images, and MRI brain tumor segmentation images.

4.1. Application of Machine Learning in Microwave Breast Cancer Images

In recent years, researchers have begun to use ML methods to directly process the signals obtained by the microwave breast cancer detection system, and thus, achieve the purpose of breast cancer screening [131]. The method of machine learning is mainly divided into two research directions. The first one is to assume that the tumor has been confirmed to exist in the breast. According to the trained model, an evaluation of the physical characteristics is needed, such as the location and size of the tumor. The other is to use the model to determine whether the tumor exists.

Under the assumption that the tumor exists, Davis et al. used a combination of Local discriminant bases (LDB) and Principal Component Analysis (PCA) to estimate the morphological characteristics of the tumor, such as the shape and size of the tumor [132]. The detection results are shown in Figure 17. Under the condition of a 10 dB signal-to-noise ratio, the method can reach an accuracy of 97% for breast cancer detection, and the shape accuracy is 70%. Conceicao et al. took the shape of the tumor as a feature [133]. The features combined with support vector machine (SVM) are used to determine whether it is a malignant tumor. They first constructed a breast prosthesis and tumors of different shapes, then placed tumors of different shapes on the breast prosthesis for measurement to obtain training samples. After that, the PCA was used to reduce the dimensionality and feature extraction of the acquired data. Finally, they analyzed the data by an SVM classifier to establish the model.

On the assumption that the tumor cannot be determined to exist, Alshehri et al. first used Discrete Cosine Transform (DCT) to extract the characteristics of the received signal of the microwave antenna [134]. Then, an artificial neural network was used as a model for judging whether the tumor was presented. They verified the effectiveness of the algorithm in uniform and uneven breast implants. The results show that the tumor existence, size, and location detection rates for both cases are highly satisfactory, which are approximately: (i) 100%, 95.8%, and 94.3% for homogeneous cases and (ii) 100%, 93.4%, and 93.1% for heterogeneous cases, respectively. Byrne et al. used PCA to extract features of the received signal on the digital breast model. They selected the SVM as classifier on the signal of each channel [135]. The final decision result was voted by the classifiers of all channels according to the majority principle. The investigated method can detect tumors as small as 4 mm in diameter with an accuracy over 71% in a dielectrically heterogeneous breast. In 2014, Santorelli et al. used SVM linear discriminant analysis to classify breast cancer tumors, and achieved an accuracy of 76.71% for identifying signals from the tumorous phantoms [136]. The above studies have confirmed that ML technology can efficiently assist doctors in disease diagnosis. For one thing, it can reduce the burden on doctors; for another, it can reduce the subjectivity of diagnosis, thereby making the diagnosis more objective and accurate.

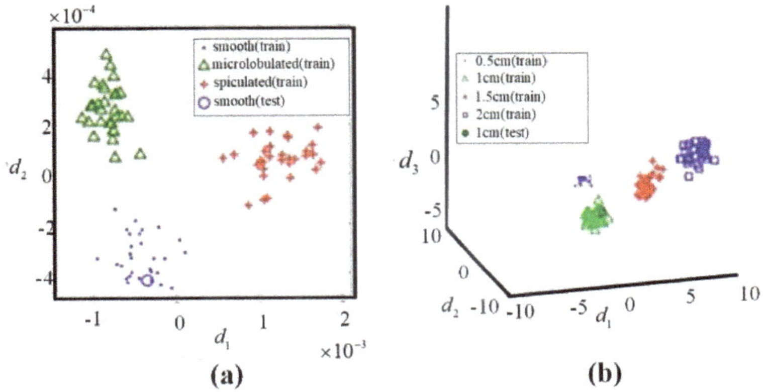

Figure 17. The classification results of the morphological characteristics of a breast tumor. (**a**) Shape. (**b**) Size.

4.2. Application of Machine Learning in MRI Prostate Cancer Images

Prostate cancer (PCa) is the cancer with the highest incidence among men. It ranks as the second-leading cause of cancer death in men. PCa has an insidious onset, but the disease progresses rapidly. The tumor cells are prone to local infiltration or distant metastasis. Therefore, early diagnosis and timely treatment have more positive meanings for PCa patients [137]. MRI is a mature and non-invasive imaging technique. Relying on the good display ability of prostate tissue structure and surrounding tissues, it occupies an important position in the diagnosis of PCa. Machine learning technology is also used in prostate cancer MRI image monitoring, mainly including prostate segmentation and cancer detection.

The segmentation of the prostate is usually a necessary step for clinical setting and further image analysis. In order to save time and increase reproducibility, it is very meaningful to realize automated prostate contour and lesion segmentation. Perfect automatic segmentation can get a fully automatic post-processing pipeline. Yan et al. proposed a model based on deep propagation neuron network by extracting the data from different levels of complexity, which can obtain more credible segmentation results of glands and their boundaries as shown in Figure 18 [138]. In addition, Alkadi et al. applied a system based on unimodal deep learning. The system can automatically segment prostate and PCa lesions [139]. Their algorithm has high segmentation accuracy and can obtain an area under the curve (AUC) of 0.995.

In terms of prostate cancer detection, ML-based computer-aided diagnosis software has important applications in the PCa field. It can improve diagnosis accuracy and repeatability. McGarry et al. used ML to detect epithelial cells and luminal density [140]. The data of 10 patients remained stable, which means the method can be used to diagnose high-grade PCa. In addition, volume of index lesion ROI can evaluate histogram parameters obtained from T2WI, DWI, and DCE images. The combination of ROI and SVM can improve the PI-RAD Sv2 score.

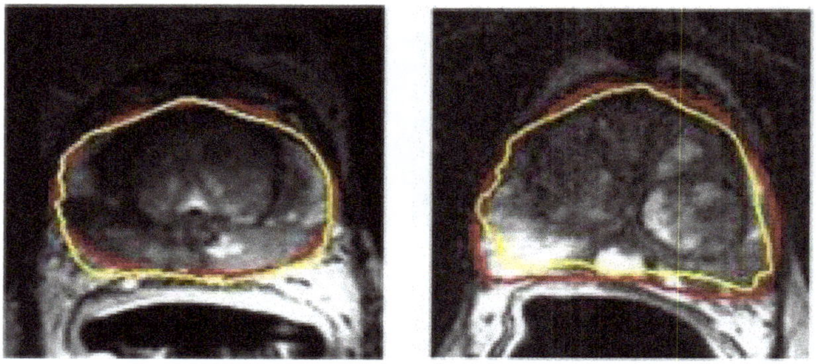

Figure 18. Segmentation results of glands and their boundaries. Red: ground truth; yellow: segmentation contour by the proposed algorithm. Adapted with permission from ref. [138]. Copyright 2018 Elsevier.

4.3. Application of Machine Learning in MRI Brain Tumor Segmentation Image

As a typical non-invasive imaging technology, magnetic resonance imaging (MRI) can produce high-quality brain images without damage and skull artifacts. MRI can provide more comprehensive information for the diagnosis and treatment of brain tumors [141]. Doctors can quantitatively analyze brain tumors to measure the maximum diameter with the help of multi-modal brain imaging to segment tumors and determine the volume and number of brain lesions. MRI brain tumor segmentation has also become an important component in the field of medical imaging because of the needs of clinical application and scientific research [142]. Researchers have proposed many methods for the segmentation of brain tumors in MRI images, but the existing methods still cannot always achieve satisfactory results due to the complexity of MRI brain images. The deep neural network model has been successfully applied to many computer vision tasks, which has received extensive attention from the scholars [143]. In specific implementation, the deep learning method segmented a large number of small-scale image blocks from the original MRI brain image. Furthermore, the image blocks are used to train a classification network. During segmentation, the corresponding image blocks are classified pixel by pixel in the context of a sliding window, and then post-processing is used to complete the segmentation of the brain tumor. This method obtains a large number of re-labeled small image blocks from the limited labeled brain image samples. Therefore, it can solve the contradiction between the few brain tumor-labeled images and the large demand for deep neural network training samples to a certain extent. The basic framework of the model for MRI brain tumor segmentation can be described as in Figure 19. The whole network architecture consisted by convolution, activation, pooling, fully connected layers, and post-processing, including conditional random fields (CRF), long short-term memory (LSTM), etc. In view of the powerful ability of automatic extraction of high discriminative features, it is quickly applied to MRI brain tumor segmentation.

The structure of a single network is used for brain tumor segmentation. The input data are processed by convolution, pooling, and nonlinear layers to obtain the feature map. Subsequently, the fully connected layer and classification layer are used to predict the label of the corresponding category based on the extracted feature. Single-network brain tumor segmentation methods can be divided into 2D and 3D due to convolution dimensions. Table 4 lists the performance of some typical single-network MRI brain tumor segmentation methods on the BraTS dataset.

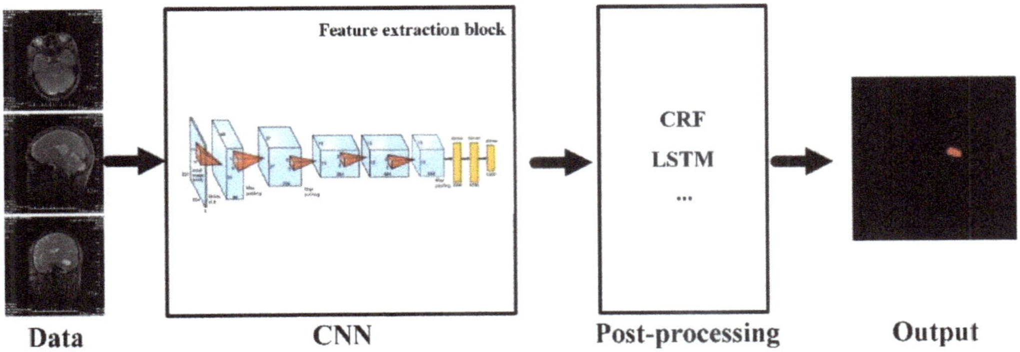

Figure 19. The basic method of MRI brain tumor segmentation based on CNN.

The 2D convolutional neural network has the advantages of occupying less resources and fast training for brain tumor segmentation. Some researchers have carried out 2D network brain tumor segmentation research. For example, Zikic et al. proposed a 2D brain tumor segmentation network by using the Alex-Net network structure [144]. Fereira et al. used a similar VGG-Net network structure to construct an automatic segmentation method for exploring the advantages of small 3 × 3 convolution kernels [35]. In addition, they explored that small convolution kernels can design deeper network architectures, which can effectively avoid overfitting by reducing the number of parameters. Dvorak and Menze proposed a local structure prediction method to improve the local segmentation ability of MRI brain images [145]. Randhawa et al. improved boundary classification of MRI brain images [146]. In order to design a more effective brain tumor segmentation model, Ben Naceur et al. adopted the idea of multi-network ensemble learning to achieve accurate brain tumor segmentation [147]. The segmentation results are shown in Figure 20. Cui et al. focused on dividing the image block into 32 × 32 or 13 × 13-pixel scales according to the different label, so as to achieve better segmentation accuracy [148]. This method improves the accuracy of brain tumor segmentation and lays the foundation for a series of later methods.

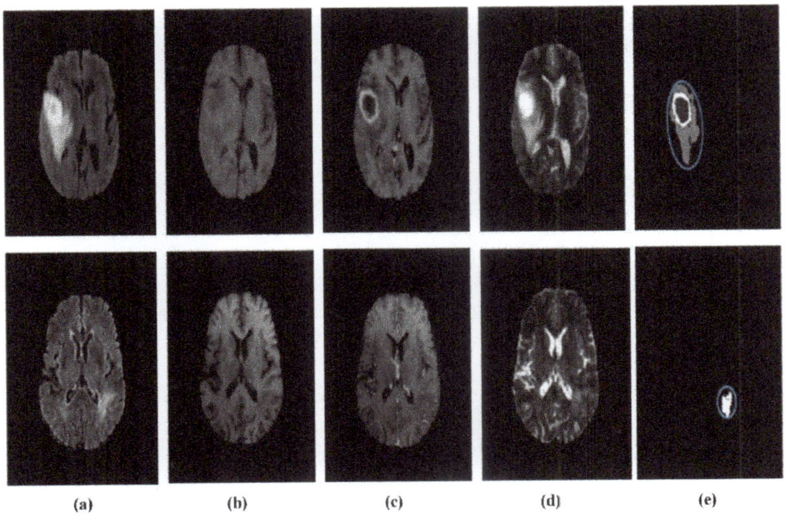

Figure 20. Brain tumor segmentations by multi-network. (**a**) Flair. (**b**) T1. (**c**) T1c. (**d**) T2. (**e**) Segmentation results. Adapted with permission from ref. [147]. Copyright 2018 Elsevier.

Table 4. Evaluation results of MRI brain tumor segmentation based on single CNN network.

Reference	Dataset	Evaluation Index DSC		
		Intact Tumor	**Core Tumor**	**Enhance Tumor**
[141]	BraTS 2013	0.88	0.83	0.77
[35]	BraTS 2013	0.86	0.75	0.73
[145]	BraTS 2014	0.83	0.75	0.77
[146]	BraTS 2016	0.87	0.75	0.71
[147]	BraTS 2017	0.89	0.76	0.81

Compared with the 2D network model, the 3D network can capture more characteristic information. The related research results are shown in Table 5. Kamnitsas et al. and Casamitjana et al. used a combination of the 3D convolution kernel and dual-pass model to improve the ability of the network to capture 3D features in MRI brain images. Among them, Kamnitsas et al. constructed a dense CNN segmentation architecture based on 3D local blocks [149]. As shown in Figure 21, the method achieved a high segmentation accuracy for brain tumor image. The proposed network combines multi-scale information by using different input sizes for each path. Casamitjana et al. employed the same input size for the two paths so that different receiving areas can focus on different information [150]. For the purpose of further improving performance, advanced network, such as Resnet and DenseNet, are gradually being introduced. Meanwhile, 3D convolution and dual-path modes are gradually being adopted in combination. Kamnitsas et al., (2017) combined dual path and residual connection blocks to construct a 3D dual-path residual segmentation network for brain tumor segmentation [151]. It can obtain better segmentation performance while using less training data and fewer filters. Chen et al. proposed a new structure that can hierarchically segment necrotic and non-enhanced tumors and edema around tumors [152]. Therefore, 3D-CNN can accurately classify each pixel in MRI brain images.

Table 5. Evaluation results of MRI brain tumor segmentation based on the multi-CNN network.

Reference	Dataset	Evaluation Index DSC		
		Intact Tumor	**Core Tumor**	**Enhance Tumor**
[150]	BraTS 2015	0.92	0.84	0.77
[151]	BraTS 2015	0.90	0.76	0.73
[152]	BraTS 2017	0.72	0.83	0.81
[153]	BraTS 2013	0.80	0.67	0.85
[154]	BraTS 2013	0.88	0.79	0.83

The generative adversarial network GAN was proposed by Google Brain researcher Goodfellow and others. It can be considered as the most representative unsupervised generative method in recent years [155]. As shown in Figure 22, the generative confrontation network consists of generator G, discriminator D, and real data X. The generator is used to generate approximately realistic objects, so that it is difficult to distinguish true from false. The discriminator is mainly used to identify the difference between the real object and the false object produced by the generator. In the training process, the generator and the discriminator overlap each other, and continuously optimize and produce the best results in the confrontation. GAN has the inherent advantages of finding the data distribution of the model by paying attention to the potential probability density of the data, which overcomes the shortcomings of relying on large amounts of data. As the most creative unsupervised generative network, the GAN has gained wide attention in many fields such as image classification and segmentation. In recent years, it has also been effectively migrated to medical image analysis tasks, including MRI brain tumor segmentation. Table 6 is the performance of the MRI brain tumor segmentation method based on the GAN.

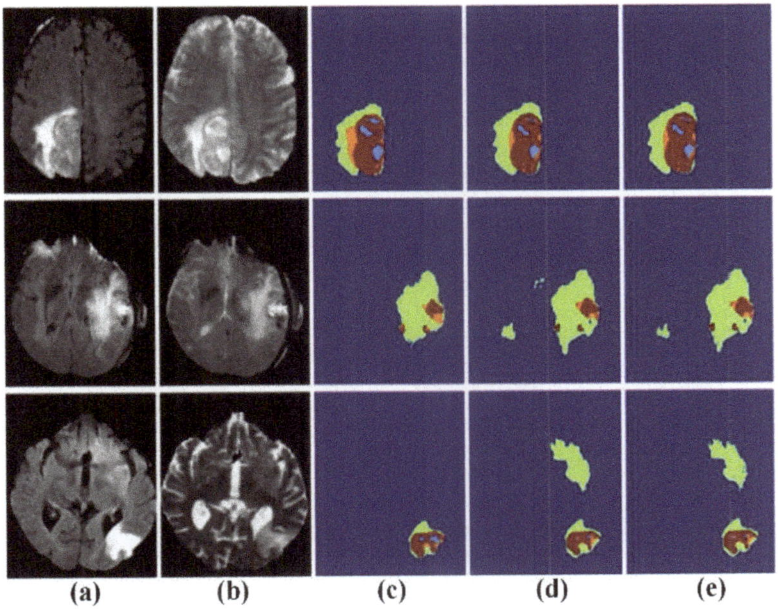

Figure 21. Brain tumor segmentations by 3D CNN. (**a**) Flair. (**b**) T2. (**c**) Manual. (**d**) Deep Medic. (**e**) Deep Medic + CRF.

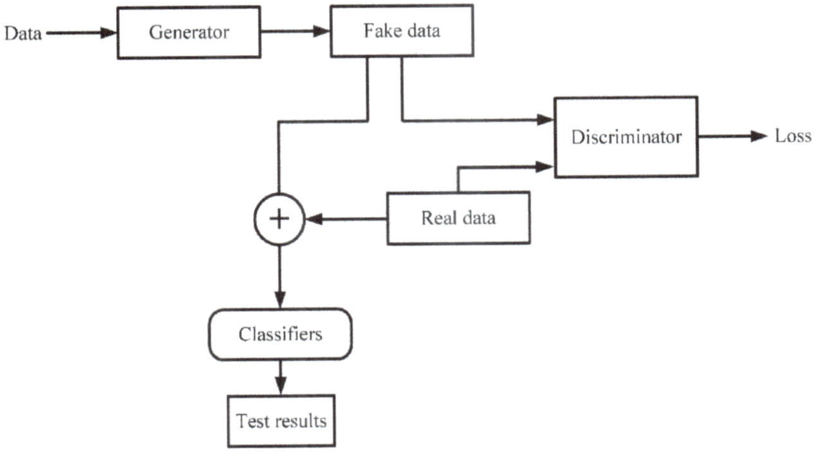

Figure 22. The MRI brain tumor segmentation based on GAN.

At the end of 2016, the semantic segmentation model based on adversarial networks is proposed. The method treats the segmentation network as a generator to train the network [156]. Li et al. tried to use GAN for MRI brain image segmentation at the MICCAI conference [157]. They trained the generator network and the discriminator network together to reduce the difference between the synthetic label and the real label. Furthermore, the high-order loss terms are used to enhance the spatial continuity of the segmentation results. Rezaei et al. proposed a new end-to-end training conditional generation confrontation network CGAN based on the GAN network [158]. The segmentation network and the discriminator network are changed to U-Net and Markovian GAN, respectively.

The two networks are simultaneously trained through backpropagation to improve the segmentation accuracy. In 2019, Rezaei et al. proposed a new confrontation network called voxel-GAN [159]. It is mainly used to alleviate the problem of data imbalance in the brain tumors segmentation, since the proportion of tumors or unhealthy areas is small. In addition, there are methods that also introduce a 3D conditional generation confrontation network to alleviate the problem of brain tumor data imbalance based on the weighted adversarial loss generated by network training.

Table 6. Evaluation results of MRI brain tumor segmentation based on GAN network.

Reference	Dataset	Evaluation Index DSC		
		Intact Tumor	Core Tumor	Enhance Tumor
[143]	BraTS 2015	0.85	0.70	0.66
[157]	BraTS 2017	0.70	0.55	0.40
[158]	BraTS 2017	0.87	0.72	0.68
[159]	BraTS 2018	0.84	0.79	0.63

5. Future Research Directions

Currently, the stability and reproducibility of biological detection are still poor due to the shortcomings of instability and variability of biologically active units. Nowadays, biological detection will inevitably be greatly developed with the rapid development of biology, informatics, materials science, and microelectronics [160–164]. It is foreseeable that electromagnetic detection will have the following characteristics in the future:

1. Integration and intelligentization

In the future, electromagnetic detection will realize information storage and memory, logical judgment, two-way communication, self-check, self-calibration, self-compensation, numerical processing, and other functions. The sensor measurement unit can become a standard unit, which can increase the portability and interchangeability of the detection system. In addition, electromagnetic detection can also be closely integrated with computers. Machine learning technology automatically analyzes the data collected by the sensor and provides analysis results more quickly and accurately, thus forming a full automation of analysis and detection.

2. Multi-sensor fusion technology

Electromagnetic detection technology can be combined with other detection technology, such as capacitive detection technology, ultrasonic detection technology, etc. The fusion data processing methods is applied to improve measurement performance. It can integrate electrical, magnetic, acoustic, thermal, force, and other multi-physical field measurements. Based on the redundant data between multiple sensors, the stability and robustness of the system can be enhanced, and the performance and stability of the system can be improved.

3. Portability

At present, flexible sensing technology has made considerable progress based on the development of soft materials, flexible electronics, and nanotechnology. The combination of electromagnetic sensors and flexible sensing technology is conducive to online and real-time monitoring. People can use portable electromagnetic sensors for daily autonomous measurement, making it possible for people to diagnose diseases and monitor food quality in daily life.

4. Mechanism Research

At present, there is no reasonable explanation for the mechanism of some electromagnetic biological effects. Mechanism research should focus on reflecting biological issues from the perspective of physics. The establishment of standardized experimental models is particularly important for mechanism research. At the same time, new theories and methods learned from the fields of mathematics, physics, chemistry, and life sciences are used to achieve a major breakthrough in the study of biological electromagnetic mechanism. Based on the electromagnetic biological effect mechanism, the application range of electromagnetic detection can be further expanded. More and more parameter information can be obtained according to the measured signal inversion. More important, the study of mechanism also contributes to the improvement of sensor sensitivity and stability.

5. Networked detection

With the rapid development of network communication technology, networked detection systems that link distributed sensor nodes and fusion centers through communication networks have gradually received widespread attention. The networked sensor system improves the flexibility and scalability of system design due to its advantages, such as less wiring, low cost, easy expansion, and maintenance. The biosensor integrates a wireless transmitter chip, which can automatically transmit the collected data to the server and display it on the mobile terminal in real time. People can check biological information anytime and anywhere to facilitate daily life. Therefore, sensor networking is also an important development direction with the application of the Internet.

6. Conclusions

Electromagnetic detection technology has developed rapidly during the 40-year development process, which has attracted an increasing number of scholars. Since bioelectromagnetic detection has different mechanisms at different frequency ranges, they are widely used in disease diagnosis and treatment, food monitoring, environmental testing, etc. In particularly, the development of MRI technology has pushed electromagnetic detection to its peak. Therefore, this review summarizes the electromagnetic biological effects and their mechanism of action. Considering the excitation frequency and intensity significantly affects the electromagnetic biological effects; this review takes the excitation frequency of the electromagnetic field as a clue to introduce the application of bioelectromagnetic detection. Moreover, electromagnetic detection in combination with ML technology has been applied in clinical diagnosis, since ML can extract the deep information of data. The research involving the application of ML technology to electromagnetic medical images is also discussed. Finally, we present an opinion on the future development trend of electromagnetic detection and make a conclusion on the reviewed biosensing techniques.

Author Contributions: P.H. prepared and wrote the manuscript; L.X. edited the manuscript; Y.X. reviewed the manuscript and prepared the figure/table configuration and literature research. All authors have read and agreed to the published version of the manuscript.

Funding: Yuedong Xie received the National Natural Science Foundation of the China Youth Fund Project (61901022) as well as Fundamental Scientific Research Business Fees of the Central Universities-Top-notch Young Talents Program (YWF-19-BJ-J-358).

Institutional Review Board Statement: Not applicable.

Informed Consent Statement: Not applicable.

Data Availability Statement: Not applicable.

Conflicts of Interest: The authors declare no conflict of interest.

References

1. Abdulazeez, A.T. Progress in utilisation of graphene for electrochemical biosensors. *Biosens. Bioelectron.* **2018**, *106*, 149–178.
2. Zhang, C.; Song, H.Z.; Guo, W.L.; Wu, H.; Xu, X.; Yan, S.C. Multi-index detection electrochemical biosensor based on graphene aerogel/platinum nanoparticle hybrid materials. *J. Bionanosci.* **2016**, *10*, 495–500. [CrossRef]
3. Chen, Y.T.; Lee, Y.C.; Lai, Y.H.; Lim, J.C.; Huang, N.T.; Lin, C.T.; Huang, J.J. Review of integrated optical biosensors for point-of-care applications. *Biosensors* **2020**, *10*, 209. [CrossRef] [PubMed]
4. Martynko, E.; Kirsanov, D. Application of chemometrics in biosensing: A brief review. *Biosensors* **2020**, *10*, 100. [CrossRef] [PubMed]
5. Clements, J.C.; Comeau, L.A. Use of High-frequency noninvasive electromagnetic biosensors to detect ocean acidification effects on shellfish behavior. *J. Shellfish Res.* **2019**, *38*, 811–818. [CrossRef]

6. Hwang, J.; Choi, Y.; Lee, K.; Krishnan, V.; Pelled, G.; Gilad, A.A.; Choi, J. Regulation of electromagnetic perceptive gene using ferromagnetic particles for the external control of calcium ion transport. *Biomolecules* **2020**, *10*, 308. [CrossRef] [PubMed]
7. Liu, Y.; Edge, R.; Henbest, K.; Timmel, C.R.; Hore, P.J.; Gast, P. Magnetic field effect on singlet oxygen production in a biochemical system. *Chem. Commun.* **2005**, *2*, 174–176. [CrossRef] [PubMed]
8. Pauzaite, G.; Malakauskiene, A.; Nauciene, Z.; Zukiene, R.; Filatova, I.; Lyushkevich, V.; Azarko, I.; Mildaziene, V. Changes in Norway spruce germination and growth induced by pre-sowing seed treatment with cold plasma and electromagnetic field: Short-term versus long-term effects. *Plasma Process. Polym.* **2018**, *15*, e1700068. [CrossRef]
9. Wang, H.W.; Fu, Z.H.; Wang, Y.; Tai, H.M.; Chen, W.L. On-site calibration of air-coil sensor for transient electromagnetic exploration. *Geophys. Prospect.* **2019**, *67*, 1595–1610. [CrossRef]
10. McCully, K.S. Environmental pollution, oxidative stress and thioretinaco ozonide: Effects of glyphosate, fluoride and electromagnetic fields on mitochondrial dysfunction in carcinogenesis, atherogenesis and aging. *Ann. Clin. Lab. Sci.* **2020**, *50*, 408–411. [PubMed]
11. Wang, X. Study on biological effect protection of electromagnetic radiation shielding fabric. *Adv. Mater. Res.* **2011**, *1445*, 999–1002. [CrossRef]
12. Kalantaryan, V.P.; Hakobyan, S.N.; Vardevanyan, P.O. Effect of weak electromagnetic waves on thermal properties of biomacromolecule water solutions. *J. Contemp. Phys.* **2018**, *53*, 175–178. [CrossRef]
13. Vanderstraeten, J.; Verschaeve, L. Biological effects of radiofrequency fields: Testing a paradigm shift in dosimetry. *Environ. Res.* **2020**, *184*, 109387. [CrossRef] [PubMed]
14. David, S.; Meike, M. Manmade electromagnetic fields and oxidative stress—Biological effects and consequences for health. *Int. J. Mol. Sci.* **2021**, *22*, 3772. [CrossRef]
15. Ince-Yilmaz, O.; Erol, T.; Kara, K.; Dogan, M.; Erdem, U. Impacts of high voltage electric field (HVEF) applications on germination and seedling growth of seed (*Tritcum aestivum* L.) with analysis by fourier transform infrared (FTIR) Spectroscopy. *Fresenius Environ. Bull.* **2018**, *27*, 5153–5162.
16. Li, D.P.; Jia, S.L.; Zhang, L.T.; Wang, Z.Y.; Pan, J.F.; Zhu, B.W.; Luo, Y.K. Effect of using a high voltage electrostatic field on microbial communities, degradation of adenosine triphosphate, and water loss when thawing lightly-salted, frozen common carp (*Cyprinus carpio*). *J. Food Eng.* **2017**, *212*, 226–233. [CrossRef]
17. Han, Z.W.; Fu, J.; Fang, Y.Q.; Zhang, J.Q.; Niu, S.C.; Ren, L.Q. Anti-adhesive property of maize leaf surface related with temperature and humidity. *J. Bionic Eng.* **2017**, *14*, 540–548. [CrossRef]
18. Sabouni, A.; Honrath, M.; Khamechi, M. Thermal effects in the brain during transcranial magnetic stimulation. *IEEE Magn. Lett.* **2017**, *8*, 1509603. [CrossRef]
19. Van den Brink, J.S. Thermal effects associated with RF exposures in diagnostic MRI: Overview of existing and emerging concepts of protection. *Concepts Magn. Reson. Part B* **2019**, *2019*, 9618680. [CrossRef]
20. Campi, T.; Cruciani, S.; De Santis, V.; Palandrani, F.; Maradei, F.; Feliziani, M. Induced effects in a pacemaker equipped with a wireless power transfer charging system. *IEEE Trans. Magn.* **2017**, *53*, 9401704. [CrossRef]
21. Sehatbakhsh, S.; Kushnir, A.; Kabach, M.; Kolek, M.; Chait, R.; Ghumman, W. A case of electromagnetic interference between HeartMate 3 LVAD and implantable cardioverter defibrillator. *Pace-Pacing Clin. Electrophysiol.* **2018**, *41*, 218–220. [CrossRef]
22. Berkelmann, L.; Bader, A.; Meshksar, S.; Dierks, A.; Majernik, G.H.; Krauss, J.K.; Schwabe, K.; Manteyffel, D.; Ngezahayo, A. Tumour-treating fields (TTFields): Investigations on the mechanism of action by electromagnetic exposure of cells in telophase/cytokinesis. *Sci. Rep.* **2019**, *9*, 7362. [CrossRef] [PubMed]
23. Raco, V.; Bauer, R.; Norim, S.; Gharabaghi, A. Cumulative effects of single TMS pulses during beta-tACS are stimulation intensity-dependent. *Brain Stimul.* **2017**, *10*, 1055–1060. [CrossRef]
24. Bilgin, H.M.; Celik, F.; Gem, M.; Akpolat, V.; Yildiz, I.; Ekinci, A.; Ozerdem, M.S.; Tunik, S. Effects of local vibration and pulsed electromagnetic field on bone fracture: A comparative study. *Bioelectromagnetics* **2017**, *38*, 339–348. [CrossRef] [PubMed]
25. Israel, M.; Zaryabova, V.; Ivanova, M. Electromagnetic field occupational exposure: Non-thermal vs. thermal effects. *Electromagn. Biol. Med.* **2013**, *32*, 145–154. [CrossRef] [PubMed]
26. Jukka, J.; Anne, H.; Timo, T.; Jonne, N. Review of possible modulation-dependent biological effects of radiofrequency fields. *Bioelectromagnetics* **2011**, *32*, 511–534.
27. Goyal, R.; Vafai, K. Electromagnetic field-induced thermal management of biological materials. *Numer. Heat Transf. Part A Appl.* **2017**, *72*, 275–290. [CrossRef]
28. Donald, I. Biological effects of microwave radiation. *J. Air Pollut. Control Assoc.* **2012**, *24*, 122–127.
29. Frohlich, H. What are non-thermal electric biological effects. *Bioelectromanetics* **1982**, *3*, 45–46. [CrossRef]
30. Geesink, J.H.; Meijer, D.K.F. Bio-solition model that predicts non-thermal electromagnetic frequency bands, that either stabilize or destabilize living cells. *Electromagn. Biol. Med.* **2017**, *36*, 357–378. [CrossRef]
31. Lebedeva, N.N.; Sulimov, A.V.; Sulimova, O.P.; Kotrovskaya, T.I.; Gailus, T. Cellular phone electromagnetic field effects on bioelectric activity of human brain. *Crit. Rev. Biomed. Eng.* **2000**, *28*, 323–337. [CrossRef]
32. Ritz, T.; Adem, S.; Schulten, K. A model for photoreceptor based magnetoreception in birds. *Biophys. J.* **2000**, *78*, 707–718. [CrossRef]
33. Ritz, T.; Thalau, P.; Phillips, J.B.; Wiltschko, R.; Wiltschko, W. Resonance effects indicate a radical-pair mechanism for avian magnetic compass. *Nature* **2004**, *429*, 177–180. [CrossRef] [PubMed]
34. Lokajícek, M.; Kozubek, S.; Prokěs, K. A new concept of cumulative biological effect. *Strahlentherapie* **1981**, *157*, 41–45. [PubMed]

35. Pereira, S.; Pinto, A.; Alves, V.; Silva, C.A. Brain tumor segmentation using convolutional neural networks in MRI images. *IEEE Trans. Med. Imaging* **2016**, *35*, 1240–1251. [CrossRef] [PubMed]
36. Li, M.; Li, B.; Zhang, W.J. Rapid and non-invasive detection and imaging of the hydrocolloid-injected prawns with low-field NMR and MRI. *Food Chem.* **2018**, *242*, 16–21. [CrossRef]
37. He, X.L.; Liu, R.; Nirasawa, S.; Zheng, D.J.; Liu, H.J. Effect of high voltage electrostatic field treatment on thawing characteristics and post-thawing quality of frozen pork tenderloin meat. *J. Food Eng.* **2013**, *115*, 245–250. [CrossRef]
38. Heieh, C.W.; Ko, W.C. Effect of high—Voltage electrostatic field on quality of carrot juice during refrigeration. *LWT-Food Sci. Technol.* **2008**, *41*, 1752–1757.
39. Wu, Y.; Zhu, J.; Li, Y.; Li, M. Effect of high-voltage electrostatic field on inorganic nitrogen uptake by cucumber plants. *Trans. Aasabe* **2016**, *59*, 25–29.
40. Krueger, A.P.; Kotaka, S.; Andriese, P.C. Some observations on the physiological effects of gaseous ions. *Int. J. Biometeorol.* **1962**, *6*, 33–48. [CrossRef]
41. Palanimuthu, V.; Rajkumar, P.; Orsat, V.; Gariepy, Y.; Raghavan, G.S.V. Improving cranberry shelf-life using high voltage electric field treatment. *J. Food Eng.* **2009**, *90*, 365–371. [CrossRef]
42. Sidaway, G. Influence of electrostatic fields on seed germination. *Nature* **1966**, *211*, 303. [CrossRef]
43. Dhayal, M.; Lee, S.Y.; Park, S.U. Using low-pressure plasma for *Carthamus tinctorium* L. seed surface modification. *Vacuum* **2006**, *80*, 499–506. [CrossRef]
44. Huang, R.; Sukprakarn, S.; Phavaphutanon, L. A comparison of electric field treatments to hydropriming on cucumber seed germination enhancement. *Nat. Sci.* **2006**, *40*, 559–565.
45. Li, M.Q.; Wu, Y.Y.; Zhang, M.M.; Zhu, J.Y. High-voltage electrostatic fields increase nitrogen uptake and improve growth of tomato seedlings. *Can. J. Plant Sci.* **2018**, *98*, 93–106. [CrossRef]
46. Shao, C.Y.; Wang, D.C.; Tang, X.; Zhao, L.J.; Li, Y. Stimulating effects of magnetized arc plasma of different intensities on the germination of old spinach seeds. *Math. Comput. Model.* **2013**, *58*, 808–812. [CrossRef]
47. Xu, W.Q.; Song, Z.Q.; Luan, X.Y.; Ding, C.J.; Cao, Z.Y.; Ma, X.H. Biological effects of high-voltage electric field treatment of naked oat seeds. *Appl. Sci.* **2019**, *9*, 3829. [CrossRef]
48. Kiatgamjorn, P.; Khan-ngern, W.; Nitta, S. The comparison of electric field intensity affects to the bean sprouts growing. In Proceedings of the Asia-Pacific Conference on Environmental Electromagnetics, Hangzhou, China, 4–7 November 2003; pp. 142–147.
49. Isobe, S.; Ishida, N.; Koizumi, M.; Kano, H.; Hazlewood, C.F. Effect of electric field on physical states of cell-associated water in germinating morning glory seeds observed by 1 H-NMR. *Bichim. Biophys. Acta. Gen. Subj.* **1999**, *1426*, 17–31. [CrossRef]
50. Li, F.F.; Wang, B.; Liu, Q.; Chen, Q.; Zhang, H.W.; Xia, X.F.; Kong, B.H. Changes in myofibrillar protein gel quality of porcine longissimus muscle induced by its structural modification under different thawing methods. *Meat Sci.* **2019**, *147*, 108–115. [CrossRef]
51. Cai, L.Y.; Cao, M.J.; Regenstein, J.; Cao, A.L. Recent advances in food thawing technologies. *Compr. Rev. Food Sci. Food Saf.* **2019**, *18*, 953–970. [CrossRef]
52. Hsieh, C.W.; Lai, C.H.; Lee, C.H.; Ko, W.C. Effects of high-voltage electrostatic fields on the quality of tilapia meat during refrigeration. *J. Food Sci.* **2011**, *76*, 312–317. [CrossRef] [PubMed]
53. He, X.L.; Jia, G.L.; Tatsumi, E.; Liu, H.J. Effect of corona wind, current, electric field and energy consumption on the reduction of the thawing time during the high-voltage electrostatic-field (HVEF) treatment process. *Innov. Food Sci. Emerg. Technol.* **2016**, *34*, 135–140. [CrossRef]
54. Dalvi-Isfahan, M.; Hamdami, N.; Le-Bail, A. Effect of combined high voltage electrostatic with air blast freezing on quality attributes of lamb meat. *J. Food Process Eng.* **2018**, *41*, e12811. [CrossRef]
55. Li, D.P.; Jia, S.L.; Zhang, L.T.; Li, Q.Z.; Pan, J.F.; Zhu, B.W.; Prinyawiwatkul, W.; Luo, Y.K. Post-thawing quality changes of common carp (*Cyprinus carpio*) cubes treated by high voltage electrostatic field (HVEF) during chilled storage. *Innov. Food Sci. Emerg. Technol.* **2017**, *42*, 25–32. [CrossRef]
56. Rahbari, M.; Hamdami, N.; Mirzaei, H.; Jafari, S.M. Investigation of the histological and textural properties of chicken breast thawed by high voltage electric field. *J. Food Process Eng.* **2020**, *43*, e13543. [CrossRef]
57. Amiri, A.; Mousakhani-Ganjeh, A.; Shafiekhani, S.; Mandal, R.; Singh, A.P.; Kenari, R.E. Effect of high voltage electrostatic field thawing on the functional and physicochemical properties of myofibrillar proteins. *Innov. Food Sci. Emerg. Technol.* **2019**, *56*, 102191. [CrossRef]
58. Mousakhani-Ganjeh, A.; Hamdami, N.; Soltanizadeh, N. Impact of high voltage electric field thawing on the quality of frozen tuna fish (*Thunnus albacares*). *J. Food Eng.* **2015**, *156*, 39–44. [CrossRef]
59. Jia, G.L.; Sha, K.; Feng, X.D.; Liu, H.J. Post-thawing metabolite profile and amino acid oxidation of thawed pork tenderloin by HVEF-A short communication. *Food Chem.* **2019**, *291*, 16–21. [CrossRef]
60. Wang, S.Y.; Chen, C.T.; Yin, J.J. Effect of allyl isothiocyanate on antioxidants and fruit decay of blueberries. *Food Chem.* **2010**, *120*, 199–204. [CrossRef]
61. Zhao, R.; Hao, J.; Xue, J.; Liu, H.; Li, L. Effect of high-voltage electrostatic pretreatment on the antioxidant system in stored green mature tomatoes. *J. Sci. Food Agric.* **2011**, *91*, 1680–1686. [CrossRef] [PubMed]
62. Kao, N.Y.; Tu, Y.F.; Sridhar, K.; Tsai, P.J. Effect of a high voltage electrostatic field (HVEF) on the shelf-life of fresh-cut broccoli (*Brassica oleracea* var. *italica*). *LWT-Food Sci. Technol.* **2019**, *116*, 108532. [CrossRef]

63. Takaki, K.; Hayashi, N.; Wang, D.; Ohshima, T. High-voltage technologies for agriculture and food processing. *J. Phys. D Appl. Phys.* **2019**, *52*, 473001. [CrossRef]
64. Ko, W.C.; Shi, H.Z.; Chang, C.K.; Huang, Y.H.; Chen, Y.A.; Hsieh, C.W. Effect of adjustable parallel high voltage on biochemical indicators and actomyosin Ca^{2+}-ATPase from tilapia (*Orechromis niloticus*). *LWT-Food Sci. Technol.* **2016**, *69*, 417–423. [CrossRef]
65. Christopher, L.; Alan, A. The DNA damage response and cancer therapy. *Nature* **2012**, *481*, 287–294.
66. Kirson, E.D.; Gurvich, Z.; Schneiderman, R.; Dekel, E.; Itzhaki, A.; Wasserman, Y.; Schatzberger, R.; Palti, Y. Disruption of cancer cell replication by a alternating electric fields. *Cancer Res.* **2004**, *64*, 3288–3295. [CrossRef] [PubMed]
67. Nie, Y.Z.; Du, L.L.; Mou, Y.B.; Xu, Z.J.; Weng, L.H.; Du, Y.W.; Zhu, Y.N.; Hou, Y.Y.; Wang, T.T. Effect of low frequency magnetic fields on melanoma: Tumor inhibition and immune modulation. *BioMed Cent.* **2013**, *13*, 582. [CrossRef] [PubMed]
68. Kim, S.J.; Jang, Y.W.; Hyung, K.E.; Lee, D.K.; Hyun, K.H.; Jeong, S.H.; Min, K.H.; Kang, W.; Jeong, J.H.; Park, S.Y. Extremely Low-frequency electromagnetic field exposure enhances inflammatory response and inhibits effect of antioxidant in RAW 264.7 cells. *Bioelectromagnetics* **2017**, *38*, 374–385. [CrossRef] [PubMed]
69. Simko, M. Cell type specific redox status is responsible for diverse electromagnetic field effects. *Curr. Med. Chem.* **2007**, *14*, 1141–1152. [CrossRef]
70. Wang, T.T.; Nie, Y.Z.; Zhao, S.L.; Han, Y.; Du, Y.W.; Hou, Y.Y. Involvement of midkine expression in the inhibitory effects of low-frequency magnetic fields on cancer cells. *Bioelectromagnetics* **2011**, *32*, 443–452. [CrossRef]
71. Ren, J.; Ding, L.; Xu, Q.Y.; Shi, G.P.; Li, X.J.; Li, X.J.; Ji, J.J.; Zhang, D.Y.; Wang, Y.P.; Wang, T.T.; et al. LF-MF inhibits iron metabolism and suppresses lung cancer through activation of P53-miR-34a-E2F1/E2F3 pathway. *Sci. Rep.* **2017**, *7*, 749. [CrossRef]
72. Zhang, X.Y.; Xue, Y.; Zhang, Y. Effects of 0.4 T rotating magnetic field exposure on density, strength, calcium and metabolism of rat thigh bones. *Bioelectromagnetics* **2006**, *27*, 1–9. [CrossRef]
73. Du, S.H.; Li, J.X.; Du, C.H.; Huang, Z.M.; Chen, G.N.; Yan, W.Q. Overendocytosis of superparamagnetic iron oxide particles increases apoptosis and triggers autophagic cell death in human osteosarcoma cell under a spinning magnetic field. *Oncotarget* **2017**, *8*, 9410–9424. [CrossRef] [PubMed]
74. Cheng, Y.; Muroski, M.E.; Peteit, D.C.M.C.; Mansell, R.; Vemulkar, T.; Morshed, R.A.; Han, Y.; Balyasnikova, I.V.; Horbinski, C.M.; Huang, X.L. Rotating magnetic field induced oscillation of magnetic particles for in vivo mechanical destruction of malignant glioma. *J. Control Release* **2016**, *223*, 75–84. [CrossRef]
75. Sisken, B.F.; Kanje, M.; Lundborg, G.; Herbst, E.; Kurtz, W. Stimulation of rat sciatic nerve regeneration with pulsed electromagnetic fields. *Brain Res.* **1989**, *485*, 309–316. [CrossRef]
76. Sunkari, V.G.; Aranovitch, B.; Portwood, N.; Nikoshkov, A. Effects of a low-intensity electromagnetic field on fibroblast migration and proliferation. *Electromagn. Biol. Med.* **2011**, *30*, 80–85. [CrossRef] [PubMed]
77. Zhong, C.; Zhang, X.; Xu, Z.J.; He, R.X. Effects of low-intensity electromagnetic fields on the proliferation and differentiation of cultured mouse bone marrow stromal cells. *Phys. Ther.* **2012**, *92*, 1208–1219. [CrossRef] [PubMed]
78. Diniz, P.; Soejima, K.; Ito, G. Nitric oxide mediates the effects of pulsed electromagnetic field stimulation on the osteoblast proliferation and differentiation. *Nitric Oxide* **2002**, *7*, 18–23. [CrossRef]
79. Liu, H.X.; Lees, P.; Abbott, J.; Bee, J.A. Pulsed electromagnetic fields preserve proteoglycan composition of extracellular matrix in embryonic chick sternal cartilage. *Bichim. Biophys. Acta Gen. Subj.* **1997**, *1336*, 303–314. [CrossRef]
80. Rastogi, P.; Lee, E.G.; Hadimani, R.L.; Jiles, D.C. Transcranial magnetic stimulation: Development of a novel deep-brain triple-Halo Coil. *IEEE Magn. Lett.* **2019**, *10*, 3102205. [CrossRef]
81. George, M.S.; Taylor, J.J.; Short, E.B. The expanding evidence base for rTMS treatment of depression. *Curr. Opin. Psychiatry* **2013**, *26*, 13–18. [CrossRef]
82. Philip, N.S.; Aiken, E.E.; Kelley, M.E.; Burch, W.; Waterman, L.; Holtzheimer, P.E. Synchronized transcranial magnetic stimulation for posttraumatic stress disorder and comorbid major depression. *Brain Stimul.* **2019**, *12*, 1335–1337. [CrossRef]
83. Jiang, C.G.; Zhang, T.; Yue, F.G.; Yi, M.L.; Gao, D. Efficacy of repetitive transcranial magnetic stimulation in the treatment of patients with chronic primary insomnia. *Cell Biochem. Biophys.* **2013**, *67*, 169–173. [CrossRef] [PubMed]
84. Levitt, J.; Kalender, G.; O'neill, J. Dorsolateral prefrontal γ-aminobutyric acid in patients with treatment-resistant depression after transcranial magnetic stimulation measured with magnetic resonance spectroscopy. *J. Psychiatry Neurosci. JPN* **2019**, *44*, 386–394. [CrossRef] [PubMed]
85. Leblhuber, F.; Sreiner, K.; Fuchs, D. Treatment of patients with geriatric depression with repetitive transcranial magnetic stimulation. *J. Neural Transm.* **2019**, *126*, 1105–1110. [CrossRef] [PubMed]
86. Deng, Z.D.; Lisanby, S.H.; Peterchev, A.V. Electric field depth-focality tradeoff in transcranial magnetic stimulation: Simulation comparison of 50 coil designs. *Brain Stimul.* **2013**, *6*, 1–13. [CrossRef]
87. Zucca, M.; Bottauscio, O.; Chiampi, M.; Zilberti, L. Operator safety and field focality in aluminum shielded transcranial magnetic stimulation. *IEEE Trans. Magn.* **2017**, *53*, 5100704. [CrossRef]
88. Colak-Atmaca, M.; Ciftci-Kavalioglu, B.; Aldan, M.A.; Argun, T.; Atmaca, M.M.; Gul, G.; Soysal, A. Investigation of the effects of continuous theta burst transcranial magnetic stimulation in patients with migraine. *Neurol. Sci. Neurophysiol.* **2018**, *35*, 177–182. [CrossRef]
89. Shimizu, T.; Hosomi, K.; Maruo, T.; Goto, Y.; Yokoe, M.; Kageyama, Y.; Shimokawa, T.; Yoshimine, T.; Saitoh, Y. Efficacy of deep rTMS for neuropathic pain in the lower limbs: A randomized, double-blind crossover trial of an H-coil and figure-8 coil. *J. Neurosurg.* **2017**, *127*, 1172–1180. [CrossRef]

90. Lu, P.P.; Wang, L. Research progress of transcranial magnetic for posttraumatic stress disorder. *Prog. Biochem. Biophys.* **2020**, *47*, 319–334.
91. Lu, M.; Ueno, S. Computational study toward deep transcranial magnetic stimulation using coaxial circular coils. *IEEE Trans. BioMed Eng.* **2015**, *62*, 2911–2919. [CrossRef]
92. Samoudi, A.M.; Emmeric, T.; Luc, M.; Wout, J.; Toshiyuki, S. Deep transcranial magnetic stimulation: Improved coil design and assessment of the induced fields using MIDA model. *BioMed Res. Int.* **2018**, *2018*, 7061420. [CrossRef]
93. Wei, X.L.; Li, Y.; Lu, M.L.; Wang, J.; Yi, G.S. Comprehensive survey on improved focality and penetration depth of transcranial magnetic stimulation employing multi-coil arrays. *Int. J. Environ. Res. Public Health* **2017**, *14*, 1388. [CrossRef]
94. Li, J.T.; Cao, H.; Zhao, Z.; Zheng, M.J.; Liu, Y.H.; He, J.X.; Sun, Y.; Rem, Z.Y. A study on the multi-channel TMS device *IEEE Trans. Magn.* **2017**, *53*, 1–5. [CrossRef]
95. Sun, Z.B.; Liang, L.M.; Li, J.K.; Liu, X.Y.; Sun, J.; Zou, X.B.; Zou, M.; Guo, Z.M. Rapid detection of Atlantic salmon multi-quality based on impedance properties. *Food Sci. Nutr.* **2020**, *8*, 862–869. [CrossRef]
96. Schwan, A.; Herman, P. Electrical properties of tissue and cell suspensions. *Adv. Biol. Med. Phys.* **1957**, *5*, 147–209. [PubMed]
97. Aalhus, J.L.; Larsen, I.L.; Dubeski, P.L.; Jeremiah, L.E. Improved beef tenderness using a modified on-line carcass suspension method with or without low voltage electrical stimulation. *Can. J. Anim. Sci.* **2000**, *80*, 51–58. [CrossRef]
98. Swatland, H.J. Observations on rheological, electrical, and optical changes during rigor development in pork and beef. *J. Anim. Sci.* **1997**, *75*, 975–985. [CrossRef]
99. Ibba, P.; Falco, A.; Abera, B.D.; Cantarella, G.; Petti, L.; Lugli, P. Bio-impedance and circuit parameters: An analysis for tracking fruit ripening. *Postharvest Biol. Technol.* **2020**, *159*, 110978. [CrossRef]
100. Castro-Giraldez, M.; Botella, P.; Toldra, F.; Fito, P. Low-frequency dielectric spectrum to determine pork meat quality. *Innov. Food Sci. Emerg. Technol.* **2010**, *11*, 376–386. [CrossRef]
101. Kim, J.; Abbasi, M.A.; Kim, T.; Park, K.D.; Cho, S. Lock-in amplifier-based impedance detection of tissue type using a monopolar injection needle. *Sensors* **2019**, *19*, 4614. [CrossRef] [PubMed]
102. Tang, J.W.; Yin, W.L.; Lu, M.Y. Bio-impedance spectroscopy for frozen-thaw of bio-samples: Non-contact inductive measurement and finite element (FE) based cell modelling. *J. Food Eng.* **2020**, *272*, 109784. [CrossRef]
103. Tang, J.W.; Lu, M.Y.; Xie, Y.D.; Yin, W.L. A Novel Efficient FEM Thin shell model for bio-impedance analysis. *Biosensors* **2020**, *10*, 69. [CrossRef]
104. Yin, W.L.; Tang, J.W.; Lu, M.Y.; Xu, H.Y.; Huang, R.C.; Zhao, Q.; Zhang, Z.J.; Peyton, A. An equivalent-effect phenomenon in eddy current non-destructive testing of thin structures. *IEEE Access* **2019**, *7*, 70296–70307. [CrossRef]
105. Li, M.F.; Hansen, C.; Rose, G. A simulator for advanced analysis of a 5-DOF EM tracking systems in use for image-guided surgery. *Int. J Comput. Assist. Radiol. Surg.* **2017**, *12*, 2217–2229. [CrossRef]
106. Kim, W.; Song, J.; Park, F.C. Closed-form position and orientation estimation for a three-axis electromagnetic tracking system. *IEEE Trans. Ind. Electron.* **2018**, *65*, 4331–4337. [CrossRef]
107. Lemmers, G.P.G.; Heijmans, M.W.M.; Scafoglieri, A.; Buyl, R.; Staal, J.B.; Schmitt, M.A.; Cattrysse, E. Three-dimensional kinematics of the cervical spine using an electromagnetic tracking device. Differences between healthy subjects and subjects with nonspecific neck pain and the effect of age. *Clin. Biomech.* **2018**, *54*, 111–117. [CrossRef] [PubMed]
108. Xie, Y.D.; Lu, M.Y.; Yin, W.L.; Xu, H.Y.; Zhu, S.; Tang, J.W.; Chen, L.M.; Ran, Q.Y.; Zhang, Y.N.; Qu, Z.G. Novel wearable sensors for biomechanical movement monitoring based on electromagnetic sensing techniques. *IEEE Sens. J.* **2020**, *20*, 1019–1027. [CrossRef]
109. Robinson, D.A. A method of measuring eye movement using a scleral search coil in a magnetic field. *IEEE Trans. Biomed. Eng.* **1963**, *10*, 137–145. [PubMed]
110. Remmel, R.S. An inexpensive eye movement monitor using the scleral search coil technique. *IEEE Trans. Biomed. Eng.* **1984**, *31*, 388–390. [CrossRef]
111. Remmel, R.S. Use of an electromagnetic eye movement monitor for easy measurement of arm movements. *IEEE Trans. Biomed. Eng.* **2006**, *53*, 2356–2361. [CrossRef] [PubMed]
112. St Pierre, T.G.; Dobson, J. Theoretical evaluation of cell membrane ion channel activation by applied magnetic fields. *Eur. Biophys. J.* **2000**, *29*, 455–456. [CrossRef]
113. Ito, D.; Numano, T.; Mizuhara, K.; Takamoto, K.; Onishi, T.; Nishijo, H. A new technique for MR elastography of the supraspinatus muscle: A gradient-echo type multi-echo sequence. *Magn. Reson. Imaging* **2016**, *34*, 1181–1188. [CrossRef]
114. Ye, S.; Feng, S.L.; Huang, L.; Bian, S.T. Recent progress in wearable biosensors: From healthcare monitoring to sports analytics. *Biosensors* **2020**, *10*, 205. [CrossRef] [PubMed]
115. He, X.W.; Yuan, R.B.; Li, B.K.; Hou, Y.Q. The design of an open MRI 4-channel receive-only phased array knee coil. *Appl. Magn. Reson.* **2016**, *47*, 499–510. [CrossRef]
116. Fujita, H.; Zheng, T.H.; Yang, X.Y.; Finnerty, M.J.; Handa, S. RF surface receive array coils: The art of an LC circuit. *J. Magn. Reson. Imageing* **2013**, *38*, 12–25. [CrossRef] [PubMed]
117. Hardy, C.J.; Giaquinto, R.O.; Piel, J.E.; Rohling, K.W.; Marinelli, L.; Blezek, D.J.; Fiveland, E.W.; Darrow, R.D.; Foo, T.K.F. 128-channel body MRI with a flexible high-density receiver-coil array. *J. Mgan. Reson. Imaging* **2008**, *28*, 1219–1225. [CrossRef] [PubMed]
118. Nordmeter-Massner, J.A.; De Zanche, N.; Pruessmann, K.P. Mechanically adjustable coil array for wrist MRI. *Magn. Reson. Med.* **2009**, *61*, 429–438. [CrossRef] [PubMed]

119. Nordmeter-Massner, J.A.; De Zanche, N.; Pruessmann, K.P. Stretchable coil arrays: Application to knee imaging under varyin flexion angles. *Magn. Reson. Med.* **2012**, *67*, 872–879. [CrossRef]
120. Jia, F.; Yuan, H.Y.; Zhou, D.G.; Zhang, J.; Wang, X.Y.; Fang, J. Knee MRI under varying flexion angles utilizing a flexible flat cabl antenna. *NMR Biomed.* **2015**, *28*, 460–467. [CrossRef]
121. Corea, J.R.; Flynn, A.M.; Lechene, B.; Scott, G.; Reed, G.D.; Shin, P.J.; Lustig, M.; Arias, A.C. Screen-printed flexible MRI receiv coils. *Nat. Commun.* **2016**, *7*, 10839. [CrossRef]
122. Zidane, M.A.; Amar, H.; Rouane, A. Study of two constraints impacting measurements of human glycemia using a microwav sensor. *Biosensors* **2021**, *11*, 83. [CrossRef]
123. Wang, L.L. Microwave sensors for breast cancer detection. *Sensors* **2018**, *18*, 655. [CrossRef] [PubMed]
124. Bassi, M.; Caruso, M.; Khan, M.S.; Bevilacqua, A.; Capobianco, A.D.; Neviani, A. An integrated microwave imaging radar wit planar antennas for breast cancer detection. *IEEE Trans. Microw. Theory* **2013**, *61*, 2108–2118. [CrossRef]
125. Yang, Z.K.; Cheng, C.C.; Wang, Z.B.; Yang, H.W. Detection of breast cancer using ultra-wide band beamforming algorithm *Mod. Phys. Lett. B* **2017**, *31*, 1750091. [CrossRef]
126. Rappaport, C. A dispersive microwave model for human breast tissue suitable for FDTD computation. *IEEE Antennas Wire Propag. Lett.* **2007**, *6*, 179–181. [CrossRef]
127. Singer, C.F.; Tan, Y.Y.; Muhr, D.; Rappaport, C.; Gschwantler-Kaulich, D.; Grimm, C.; Polterauer, S.; Pfeiler, G.; Berger, A.; Tea M.K.M. Association between family history, mutation locations, and prevalence of BRCA1 or 2 mutations in ovarian cance patients. *Cancer Med.* **2019**, *8*, 1875–1881. [CrossRef]
128. Mohammed, B.; Abbosh, A.; Ireland, D. Circular antenna array for brain imaging systems. In Proceedings of the IEEE Internationa Symposium on Antennas and Propagation, Chicago, IL, USA, 8–14 July 2012; pp. 1125–1126.
129. Ireland, D.; Bialkowski, K.; Abbosh, A. Microwave imaging for brain stroke detection using Born iterative method. *IET Microw Antennas Propag.* **2013**, *7*, 909–915. [CrossRef]
130. Li, Y.; Zhang, X.R.; Zhang, B.; Lei, M.Y.; Cui, W.G.; Guo, Y.Z. A channel-projection mixed-scale convolutional neural network fo motor imagery EEG decoding. *IEEE Trans. Neural Syst. Rehabil. Eng.* **2019**, *27*, 1170–1180. [CrossRef]
131. Song, H.C.; Li, Y.P.; Men, A.C. Microwave breast cancer detection using time-frequency representations. *Med Biol. Eng. Comput. J Int. Fed. Med Biol. Eng.* **2018**, *56*, 571–582. [CrossRef]
132. Davis, S.K.; Van Veen, B.D.; Hagness, S.C.; Kelcz, F. Breast tumor characterization based on ultrawideband microwave backscatte *IEEE Trans. Biomed. Eng.* **2008**, *55*, 237–246. [CrossRef]
133. Conceicao, R.C.; Mediros, H.; O'Halloran, M.; Rodriguez-Herrera, D.; Flores-Tapia, D.; Pistorius, S. SVM-based classification o breast tumour phantoms using a UWB radar prototype system. In Proceedings of the URSI General Aassembly and Scientific Symposium, Beijing, China, 16–23 August 2014.
134. Alshehri, S.A.; Khatun, S.; Jantan, A.B.; Abdullah, R.S.A.R.; Mahmood, R.; Awang, Z. Experimental breast tumor detection using NN-based UWB imaging. *Prog. Electromagn. Res. PIER* **2011**, *111*, 447–465. [CrossRef]
135. Byrne, D.; Ohalloran, M.; Jones, E.; Glavin, M. Support Vector Machine-Based Ultrawideband Breast Cancer Detection System *J. Electromagn. Waves Appl.* **2011**, *25*, 1807–1816. [CrossRef]
136. Santorelli, A.; Porter, E.; Kirshin, E.; Liu, Y.J.; Popovic, M. Investigation of classifiers for tumor detection with an experimental time-domain breast screening system. *Prog. Electromagn. Res. PIER* **2014**, *144*, 45–47. [CrossRef]
137. Sun, Y.; Reynolds, H.M.; Parameswaran, B.; Wraith, D.; Finnegan, M.; Williams, S.; Haworth, A. Multiparametric MRI and radiomics in prostate cancer: A review. *Australas. Phys. Eng. Sci. Med.* **2019**, *42*, 3–25. [CrossRef] [PubMed]
138. Yan, K.; Wang, X.Y.; Kim, J.; Khadra, M.; Fulham, M.; Feng, D.G. A propagation-DNN: Deep combination learning of multi-level features for MR prostate segmentation. *Comput. Methods Programs Biomed.* **2019**, *170*, 11–21. [CrossRef]
139. Alkadi, R.; Taher, F.; El-baz, A.; Werghi, N. A deep learning-based approach for the detection and localization of prostate cancer in T2 magnetic resonance images. *J. Digit. Imaging* **2019**, *32*, 793–807. [CrossRef] [PubMed]
140. Mcgarry, S.; Hurrell, S.L.; Iczkowksi, K.A.; Hall, W.; Kaczmarowski, A.L.; Banerjee, A.; Keuter, T.; Jacobsohn, K.; Bukowy, J.D. Nevalainen, M.T.; et al. Radio-pathomic maps of epithelium and lumen density predict the location of high-grade prostate cancer *Int. J. Radiat. Oncol. Biol. Phys.* **2018**, *101*, 1179–1187. [CrossRef]
141. Esteva, A.; Kuprel, B.; Novoa, R.; Ko, J.; Swetter, S.; Blau, H.; Thrun, S. Dermatologist-level classification of skin cancer with deep neural networks. *Nature* **2017**, *542*, 115–118. [CrossRef]
142. Xue, Y.; Xu, T.; Zhang, H.; Long, L.R.; Huang, X.L. Segan: Adversarial network with multi-scale L1 loss for medical image segmentation. *Neuroinformatics* **2018**, *16*, 383–392. [CrossRef]
143. Chen, H.; Dou, Q.; Yu, L.Q.; Qin, J.; Heng, P.A. VoxRes Net: Deep voxelwise residual networks for brain segmentation from 3D MR images. *Neuroimage* **2018**, *170*, 446–455. [CrossRef]
144. Zikic, D.; Ioannou, Y.; Brown, M.; Criminisi, A. Segmentation of brain tumor tissues with convolutional neural networks. In Proceedings of the MICCAI-BRATS, Boston, MA, USA, 9 September 2014; pp. 36–39.
145. Dvorak, P.; Menze, B. Local structure prediction with convolutional neural networks for multimodal brain tumor segmentation In Proceedings of the International MICCAI Workshop on Medical Computer Vision, Munich, Germany, 9 October 2016; Springer Berlin/Heidelberg, Germany, 2015; pp. 59–71.

146. Randhawa, R.S.; Modi, A.; Jain, P.; Warier, P. Improving boundary classification for brain tumor segmentation and longitudinal disease progression. In Proceedings of the 2nd International Brainlesion: Glioma, Multiple Sclerosis, Stroke and Traumatic Brain Injuries, Athens, Greece, 9 October 2016; pp. 65–74.
147. Ben Naceur, M.; Saouli, R.; Akil, M.; Kachouri, R. Fully automatic brain tumor segmentation using end-to-end incremental deep neural networks in MRI images. *Comput. Methods Programs Biomed.* **2018**, *166*, 39–49. [CrossRef]
148. Cui, Z.P.; Yang, J.; Qiao, Y. Brain MRI segmentation with patch-based CNN approach. In Proceedings of the 35th Chinese Control Conference, Chengdu, China, 27–29 July 2016; pp. 7026–7031.
149. Kamnitsas, K.; Ferrante, E.; Parisot, S.; Ledig, C.; Nori, A.V.; Criminisi, A.; Rueckert, D.; Glocker, B. DeepMedic for brain tumor segmentation. In Proceedings of the 2nd International Brainlesion: Glioma, Multiple Sclerosis, Stroke and Traumatic Brain Injuries, Athens, Greece, 9 October 2016; pp. 138–149.
150. Casamitjana, A.; Puch, S.; Aduriz, A.; Vilaplana, V. 3D convolutional neural networks for brain tumor segmentation: A comparison of multi-resolution architectures. In Proceedings of the 2nd International Brainlesion: Glioma, Multiple Sclerosis, Stroke and Traumatic Brain Injuries, Athens, Greece, 9 October 2016; pp. 150–161.
151. Kamnitsas, K.; Ledig, C.; Newcombe, V.F.J.; Simpson, J.P.; Kane, A.D.; Menon, D.K.; Rueckert, D.; Glocker, B. Efficient multi-scale 3D CNN with fully connected CRF for accurate brain lesion segmentation. *Med. Image Anal.* **2017**, *36*, 61–78. [CrossRef] [PubMed]
152. Chen, L.L.; Wu, Y.; Dsouza, A.M.; Abidin, A.Z.; Wismüller, A.; Xu, C.L. MRI tumor segmentation with densely connected 3D CNN. In Proceedings of the Medical Imaging 2018: Image Processing, Houston, TX, USA, 11–13 February 2018.
153. Havaei, M.; Davy, A.; Warde-Farley, D.; Biard, A.; Courville, A.; Bengio, Y.; Pal, C.; Jodoin, P.M.; Larochelle, H. Brain tumor segmentation with deep neural networks. *Med. Image Anal.* **2017**, *35*, 18–31. [CrossRef]
154. Hussain, S.; Anwar, S.M.; Majid, M. Brain tumor segmentation using cascaded deep convolutional neural network. In Proceedings of the 39th Annual International Conference of the IEEE Engineering in Medicine and Biology Society, Seogwipo, Korea, 11–15 July 2017.
155. Gao, X.; Deng, F.; Yue, X.H. Data augmentation in fault diagnosis based on the Wasserstein generative adversarial network with gradient penalty. *Neurocomputing* **2020**, *396*, 487–494. [CrossRef]
156. Moreno-Barea, F.J.; Jerez, J.M.; Franco, L. Improving classification accuracy using data augmentation on small data sets. *Expert Syst. Appl.* **2020**, *161*, 113696. [CrossRef]
157. Li, Z.; Wang, Y.Y.; Yu, J.H. Brain tumor segmentation using an adversarial network. In Proceedings of the 3rd International Brainlesion: Glioma, Multiple Sclerosis, Stroke and Traumatic Brain Injuries, Quebec City, QC, Canada, 9 September 2018; pp. 123–132.
158. Rezaei, M.; Harmuth, K.; Gierke, W.; Kellermeier, T.; Fischer, M.; Yang, H.; Meinel, C. A conditional adversarial network for semantic segmentation of brain tumor. In Proceedings of the 3rd International Brainlesion: Glioma, Multiple Sclerosis, Stroke and Traumatic, Quebec City, QC, Canada, 9 September 2018; pp. 241–252.
159. Rezaei, M.; Yang, H.J.; Meinel, C. Voxel-GAN: Adversarial framework for learning imbalanced brain tumor segmentation. In Proceedings of the 4th International Brainlesion: Glioma, Multiple Sclerosis, Stroke and Traumatic Brain Injuries, Granada, Spain, 16 September 2018.
160. Yoon, J.; Shin, M.; Lim, J.; Lee, J.Y.; Choi, J.W. Recent advances in MXene nanocomposite-based biosensors. *Biosensors* **2020**, *10*, 185. [CrossRef]
161. Maddipatla, D.; Narakathu, B.B.; Atashbar, M. Recent progress in manufacturing techniques of printed and flexible sensors: A Review. *Biosensors* **2020**, *10*, 199. [CrossRef] [PubMed]
162. Hwang, E.; Song, J.S.; Zhang, J. Integration of nanomaterials and bioluminescence resonance energy transfer techniques for sensing biomolecules. *Biosensors* **2019**, *9*, 42. [CrossRef]
163. Wang, L.L. Early diagnosis of breast cancer. *Sensors* **2017**, *17*, 1572. [CrossRef] [PubMed]
164. Karamfard, S.S.; Asl, B.M. 2-stage delay-multiply-and-sum beamforming for breast cancer detection using microwave imaging. In Proceedings of the Iranian Conference on Electrical Engineering (ICEE 2017), Tehran, Iran, 2–4 May 2017; pp. 101–106.

biosensors

MDPI

Article

A Novel Efficient FEM Thin Shell Model for Bio-Impedance Analysis

Jiawei Tang [1], Mingyang Lu [1], Yuedong Xie [2,3] and Wuliang Yin [1,*]

1 School of Electrical and Electronics Engineering, The University of Manchester, Manchester M13 9PL, UK; jiawei.tang@postgrad.manchester.ac.uk (J.T.); mingyang.lu@manchester.ac.uk (M.L.)
2 School of Instrumentation and Optoelectronic Engineering, Beihang University, Beijing 100036, China; yuedongxie@buaa.edu.cn
3 Beijing Advanced Innovation Center for Big Data-based Precision Medicine, Beihang University, Beijing 100036, China
* Correspondence: wuliang.yin@manchester.ac.uk; Tel.: +44 (0) -161-306-2885

Received: 4 May 2020; Accepted: 15 June 2020; Published: 17 June 2020

Abstract: In this paper, a novel method for accelerating eddy currents calculation on a cell model using the finite element method (FEM) is presented. Due to the tiny thickness of cell membrane, a full-mesh cell model requires a large number of mesh elements and hence intensive computation resources and long time. In this paper, an acceleration method is proposed to reduce the number of mesh elements and therefore reduce the computing time. It is based on the principle of replacing the thin cell membrane with an equivalent thicker structure. The method can reduce the number of mesh elements to 23% and the computational time to 17%, with an error of less than 1%. The method was verified using 2D and 3D finite element methods and can potentially be extended to other thin shell structures. The simulation results were validated by measurement and analytical results.

Keywords: finite element method; thin shell model; β dispersion; Maxwell–Wagner effect; bio-impedance spectroscopy

---

1. Introduction

β-dispersion analysis has been widely investigated for medical [1,2] and industrial [3,4] applications. For example, it has been proposed for the detection of tumours [5,6] and cerebral stroke [7], and it has been applied to the quality inspection of food such as meat, fruit and vegetables [3]. The mechanism of β-dispersion was first described by Schwan [8,9]. β-dispersion takes place at radio frequency, mainly from hundreds of hertz to megahertz. This dispersion is caused by Maxwell–Wagner effect, an interfacial polarisation of cell membrane blocking the ion-flow of intra and extracellular dielectrics.

Analytical solutions are well developed to calculate the β dispersion [10–12]. However, the analytical solution is only designed to analyze the models with a regular shape (i.e., spherical cell model) [13,14]. In reality, most cell shapes are anomalous. The finite element method is a methodology that discretises an entire continuous domain into discrete domains. An irregular cell shape can be simulated in a high accuracy using the finite element method. Therefore, FEM is a feasible and effective way to simulate the dielectric dispersions for models with irregular shapes [15,16].

The difficulty of FEM is that the number of meshing elements is significantly large due to the tiny thickness of the cell membrane; it takes a long time to compute with numerous meshing elements. In this paper, a novel FEM acceleration method was proposed to simulate β dispersion. The acceleration method is to reduce the number of meshing elements and computing time by replacing the full-mesh cell membrane model with an equivalent reduced-mesh model. In this paper, the FEM acceleration method was used to simulate the spherical and oval cell model in two-dimensional and three-dimensional versions.

2. Materials and Methods

2.1. Single Spherical Cell Model

The full-mesh cell model is the single shell spherical model introduced by Asami [15]. The spherical cell model is shown in Figure 1. The parameters of the cell model are $k_m = 10^{-7}$ S/m, $k_a = k_c = 1$ S/m. $\varepsilon_c = \varepsilon_a = 80$, $\varepsilon_m = 5$, R = 5000 nm, $d_m = 5$nm. k_m is the conductivity of the cell membrane. k_a and k_c are the conductivity of the extracellular and intracellular fluid, respectively. ε_c and ε_a are the relative permittivity of the intracellular and extracellular fluid, respectively. ε_m is the relative permittivity of the cell membrane. R is the radius of the cell. d_m is the cell membrane thickness [10,17,18].

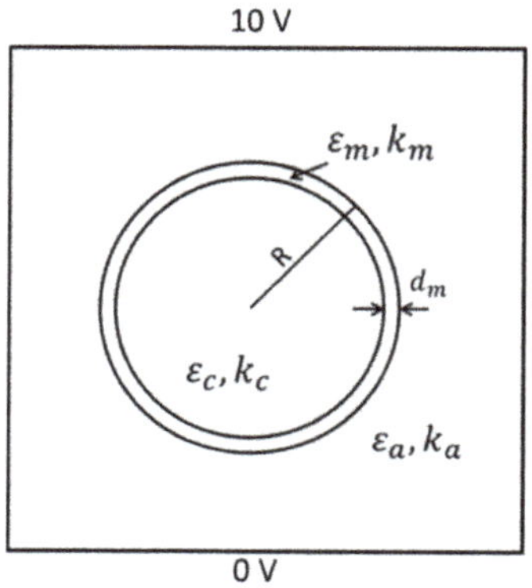

Figure 1. Spherical cell model.

The two-dimensional simulation model is a single cell model, as shown in Figure 1. The spherical cell is put in a suspension with an applied alternating electric field. The voltage of the upper electrode was set to be 10 V, and the voltage of the lower electrode was set to be 0 V [14]. The intracellular and extracellular fluid were conductive while the cell membrane blocked current flow between intracellular and extracellular fluid at low frequency. Dielectric relaxation describes a phenomenon that the permittivity decreases when the frequency of applied electric field increases at a certain range. The relaxation occurs when the electric dipoles are not able to reverse at the same rate with the frequency of the applied field. β-dispersion was observed from 10 kHz to 1 MHz. The dispersion was described by Schwan and could be explained by Maxwell–Wagner effect [8]. The effect occurs at the boundary of layered or inhomogeneous dielectrics. Applying an electric field would build up charges at the boundaries created by different dielectrics. The double-layer dielectrics system can be represented by one capacitance in parallel with one another. Applying a static electric field would charge both capacitances and creates charges at the boundary of the dielectrics. This phenomenon is also called interfacial polarisation.

The equivalent complex relative permittivity of the cell model in Figure 1 can be calculated according to Asami [12]:

$$\varepsilon_p^* = \varepsilon_m^* \frac{2(1-v)\varepsilon_m^* + (1+2v)\varepsilon_c^*}{(2+v)\varepsilon_m^* + (1-v)\varepsilon_c^*} \tag{1}$$

where $v = \left(1 - \frac{d_m}{R}\right)^n$, d_m is the cell membrane thickness, R is the radius of the cell, and n stands for the factor of dimension. When n = 2, the equation is applied to the two-dimension model. When n = 3, the equation is applied to three-dimension model. ε_p^* is the complex relative permittivity of the cell model, ε_m^* is the complex relative permittivity of the cell membrane and ε_c^* is the complex relative permittivity of cytoplasm; then the complex relative permittivity of the suspension can be calculated by Wagner's equation:

$$\varepsilon^* = \varepsilon_a^* \frac{2(1-P)\varepsilon_a^* + (1+2P)\varepsilon_p^*}{(2+P)\varepsilon_a^* + (1-P)\varepsilon_p^*} \quad (2)$$

where P is the volume fraction, ε^* is the complex relative permittivity of cell suspension, ε_a^* is the complex relative permittivity of the external medium.

The conductivity of the suspension can be derived:

$$\varepsilon^* = \frac{\sigma^*}{j\omega\varepsilon_0} = \frac{\sigma_r + j\omega\varepsilon_r\varepsilon_0}{j\omega\varepsilon_0} \quad (3)$$

where σ_r and ε_r are the conductivity and permittivity of the suspension, respectively. σ^* is the complex conductivity of the suspension, ε_0 is the permittivity of vacuum, $\omega = 2\pi f$ where f is the frequency of the applied field and j is the imaginary unit.

According to Schwan [9], when $\frac{d_m}{R} \ll 1$, $k_m \ll k_a = k_c$, the dielectric relaxation of the cell suspension can be represented by the Debye relaxation model:

$$\varepsilon^* = \varepsilon_h + \frac{\varepsilon_l - \varepsilon_h}{1 + j\omega\tau} + \frac{k_l}{j\omega\varepsilon_0} \quad (4)$$

where ε_h and ε_l are the high-frequency limit and low-frequency limit of the relative permittivity, τ is the relaxation time and k_l is the low-frequency limit of the conductivity.

2.2. D FEM Simulation

Normally, the finite element problems could be solved by two kinds of methods which are Galerkin's method and Ritz's method [19]. In this paper, Galerkin's method is used to build up simulations.

According to the Gaussian equation,

$$\nabla \cdot D = \rho \quad (5)$$

where $\vec{D} = \varepsilon\vec{E}$, $\vec{D}$ is the electric flux density, $\vec{E}$ is the electrical field intensity, $\varepsilon = \varepsilon_r\varepsilon_0$, ε_r is the relative permittivity of the material, ρ is the quantity of electric charge, which is zero in the cell model. So the equation could be rewritten as:

$$\nabla \cdot (\sigma + j\omega\varepsilon)\vec{E} = \nabla \cdot (\sigma + j\omega\varepsilon)\nabla U = 0 \quad (6)$$

where σ is the conductivity of the material and $\varepsilon = \varepsilon_r\varepsilon_0$. ε_r is the relative permittivity of the material. ε_0 is the permittivity of vacuum. U is the electric potential which is the objective value to be solved [20]. To obtain the equivalent conductivity of the total cell suspension, we calculate the total current flowing through the cell suspension from the top electrode to the bottom electrode [15]. According to Kirchhoff's current law, the current flow through any plane, which is in parallel with the electrodes is the same with the total current flow through the cell suspension. So the current I flow through a chosen plane in parallel with electrodes can be calculated as:

$$I = \sum j^{(i)} = \sum \sigma^{(i)}\vec{E}^{(i)} \quad (7)$$

where $j^{(i)}$ is the current density of the element on the chosen plane, $\sigma^{(i)}$ is the conductivity of the element on the chosen plane, $\vec{E}^{(i)}$ is the background electrical field on the chosen plane. Then the complex conductivity of the cell suspension can be calculated as:

$$\sigma^* = \frac{1}{\rho^*} = \frac{L}{S} * \frac{I}{U} \tag{8}$$

where σ^* and ρ^* are the complex conductivity and resistivity, respectively. L is the distance between the top electrode and the bottom electrode. S is the area of the electrodes.

2.3. D FEM Simulation

An induction model was built for the three-dimension FEM simulation. As shown in Figure 2, the cylinder stands for the cell suspension, and the spherical model stands for the cell. Imaginary transmitter and receiver coils are put on the top of the suspension to provide an alternating magnetic field as an exciting signal and detect induced secondary magnetic field. The single-cell model in a three-dimension situation is shown in Figure 2b. The applied field is a magnetic field excited by the transmitter in Figure 2a. The radius of the sphere cell model is R and the cell membrane thickness of the sphere cell model is d_m. All the electric parameters of the sphere cell model are set to be the same with the parameters in Figure 1. The Mxwell–Wagner effect can be calculated by calculating the eddy current distribution induced by the applied magnetic field.

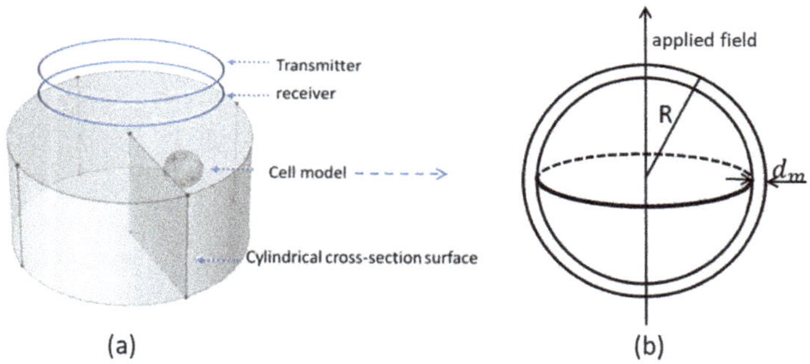

Figure 2. (**a**) 3D suspension model, (**b**) 3D single cell model.

According to Oszkar Biro [21], the basic formulas for the three-dimension edge elements simulation of eddy current problems, Galerkin's equation is shown as following:

$$\begin{aligned}&\int_{\Omega_c} \nabla \times N_i \cdot v \nabla \times A^{(n)} d\Omega + \int_{\Omega_c} j\omega\sigma N_i \cdot A^{(n)} d\Omega + \int_{\Omega_c} \sigma N_i \cdot \nabla V^{(n)} d\Omega \\ &= \int_{\Omega_c} \nabla \times N_i \cdot v_0 \nabla \times A_s d\Omega \quad i = 1,2,\ldots,6\end{aligned} \tag{9}$$

$$\int_{\Omega_c} j\omega\sigma \nabla L_i \cdot A^{(n)} d\Omega + \int_{\Omega_c} \sigma \nabla L_i \cdot \nabla V^{(n)} d\Omega = 0 \quad i = 1,2,\ldots,4 \tag{10}$$

where L_i is the elemental interpolation of ith node corresponding to the nth element of it; N_i stands for the vector interpolation of ith edge relating to the nth edge element of it; $A^{(n)}$ stands for the excited edge vector potential of the nth element; A_s represents the original edge vector potential of the nth element; $V^{(n)}$ represents the electrical potential excited by the nth element on the receiving coil; v stands for the reluctivity of the model; v_0 is the reluctivity of the air; σ is the conductivity of the model [22–24].

The eddy current density is:

$$J^{(i)} = \sigma^{(i)} E^{(i)} \tag{11}$$

where $E^{(i)}$ denotes the vector sum of the electrical field on all the edges of each tetrahedral element, $\sigma^{(i)}$ is the complex conductivity parameter of the of each tetrahedral element.

The model is shown in Figure 2. Assuming there is an equivalent model with uniform dielectric and the background suspension has exactly the same shape with the simulation model in Figure 2. Since the normal component of the electrical field relative to each cross-section surface of suspension (shown in Figure 2) is identical, then

$$\sum \vec{n}\cdot E^{(i)} = \sum \vec{n}\cdot E_s^{(i)} \tag{12}$$

where, $E_s^{(i)}$ denotes the background electrical field of the of each tetrahedral element; $\vec{n}$ is the normal unit vector relative to the surface of target;

Since the equivalent model is uniform, the electrical background field $E_s^{(i)}$ is always vertical to every cylindrical cross-section surface of the suspension model shown in Figure 2. Then the equivalent complex conductivity of the original suspension (as shown in Figure 2) can be derived from (8) and (9) that

$$\sigma_s = \frac{\sum J_s^{(i)}}{\sum \vec{n}\cdot\frac{J^{(i)}}{\sigma^{(i)}}} \tag{13}$$

$\sigma^{(i)}$ is the complex conductivity of each tetrahedral element, $J_s^{(i)}$ is the eddy current density through each cylindrical cross-section of equivalent suspension model, $J^{(i)}$ is the eddy current density through each cylindrical cross-section of the original suspension model with arranged cells and $\vec{n}$ is the current flow direction.

2.4. Acceleration Method

The complicated calculation progress and the long computing time are caused by a large number of meshing elements around the thin cell membrane. The number of meshing elements could be reduced by enlarging the cell membrane thickness. However, changing the cell membrane thickness affects the conductivity spectrum of the cell suspensions. The aim of the proposed acceleration method is to keep the accuracy of the simulation result while enlarging the cell membrane thickness (reducing the number of elements). This acceleration method replaces the full-mesh model with an equivalent thicker cell membrane (reduce-mesh model) with an equivalent complex conductivity, as shown in Figure 3. The M_1 in Figure 3 is the full-mesh model cell membrane with normal thickness l_1. The M_2 in Figure 3 is the enlarged part of the reduce-mesh model cell membrane with thickness l_2, R is the radius of the cell model. The electrical parameter of the cell membrane M_2 is exactly the same with the intracellular fluid. The cell membrane M_1 and M_2 combine as one thicker equivalent cell membrane (reduce-mesh model cell membrane), and the thickness of the reduce-mesh model cell membrane is $l_1 + l_2$. The number of meshing elements is reduced due to a thicker equivalent cell membrane. In this case, the factor $v = \left(1 - \frac{d_m}{R}\right)^n$ in Equation (1) can be rewritten as $v = \left(1 - \frac{(l_1+l_2)}{R}\right)^n$ to describe the Maxwell–Wagner effect in the equivalent model. According to the Maxwell–Wagner Equation (2), as the parameter P and ε_a^* of the equivalent model are fixed, the dielectric behaviour of the suspension model ε^* only depends on the dielectric property of the cell model ε_p^*. According to Equation (1), ε_p^* is determined by ε_m^* and the factor v. In order to retain the dielectric behaviour of the suspension model, the equivalent dielectric parameters of the equivalent cell membrane ε_m^* can be calculated by Equations (14)–(18). With the equivalent complex conductivity, the behaviour of Maxwell–Wagner effect, which leads to β-dispersion remains the same.

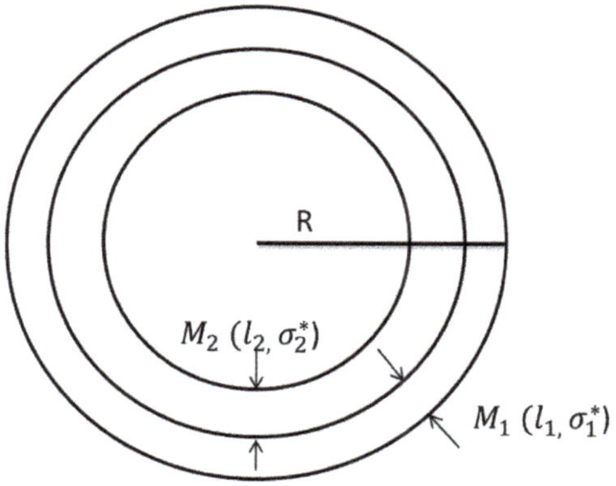

Figure 3. Equivalent cell membrane model.

The parameters of the cell membrane M_1 and M_2 in Figure 3 are:

$$\sigma_1^* = \sigma_1 + j\omega\varepsilon_1\varepsilon_0 \tag{14}$$

$$\sigma_2^* = \sigma_2 + j\omega\varepsilon_2\varepsilon_0 \tag{15}$$

where σ_1^* and σ_2^* are the complex conductivity of M_1 cell membrane and M_2 cell membrane in Figure 3, respectively. σ_1 and σ_2 are the conductivity of cell membrane M_1 and cell membrane M_2 in Figure 3, respectively. ε_1 and ε_2 are the relative permittivity of the cell membrane M_1 and cell membrane M_2, respectively. Considering the two cell membranes M_1 and M_2 are connected as two electrolytes in series. The total impedance Z is:

$$Z = Z_1 + Z_2 = \frac{l_1}{\sigma_1^* S_1} + \frac{l_2}{\sigma_2^* S_2} \tag{16}$$

where Z_1 and Z_2 are the impedance of the cell membrane M_1 and M_2, respectively. σ_1^* and σ_2^* are the complex conductivity of cell membrane M_1 and M_2 respectively. l_1 and l_2 are the thickness of the cell membrane M_1 and the cell membrane M_2 in Figure 3, respectively. S_1 and S_2 are the area of cell membrane M_1 and the cell membrane M_2, respectively. As l_1 and l_2 are negligible comparing with radius R, the area $S_1 = S_2$.

Then the equivalent complex conductivity can be derived as:

$$Z = \frac{l_1}{\sigma_1^* S_1} + \frac{l_2}{\sigma_2^* S_2} = \frac{l_1\sigma_2^* + l_2\sigma_1^*}{\sigma_1^*\sigma_2^*}\frac{1}{S_1} = \frac{(l_1 + l_2)}{\sigma_e^*}\frac{1}{S_1} \tag{17}$$

$$\sigma_e^* = \frac{(l_1 + l_2)\sigma_1^*\sigma_2^*}{l_1\sigma_2^* + l_2\sigma_1^*} \tag{18}$$

σ_e^* is the complex conductivity of the equivalent cell membrane (reduce-mesh model cell membrane).

The replacement of the thin cell membrane with an equivalent thicker structure appears to have reflected an equivalent Maxwell–Wagner effect from the computation results. Theoretically, less contrasting EM properties between the boundary of two materials cause less Maxwell–Wagner effect (i.e., less charge building-up). At the same time, thicker materials between the two charged surfaces would mean the equivalent capacitance is smaller. These two effects would mean the electric field E = Q/C would be kept to a constant; this is possibly the reason why we can use the replacement to carry out the calculation of the dispersion caused by the MW effect. Equations (17) and (18) considered the complex conductivity, and therefore, the physics would be masked to a certain degree, but the principle can be explained as above.

2.5. Meshing Details

The suspension model shown in Figure 1 was discretised into three regions, cell membrane and intra and extracellular fluid, for meshing. The commercial software Comsol 5.0 was used for meshing. The convergent criterion was accelerated BICGS with an optimal precondition. The initial precondition was based on optimised initial guess: the final solution from the previous frequency [25–30]. As shown in Table 1, the minimum and maximum element sizes are set to be the same to maintain the meshing accuracy. The maximum element growth rate controls the maximum changing rate of the dimension of the element among the adjacent subdomains. The narrow factor controls the mesh density on the narrow region. Increasing the cell membrane thickness reduces the narrow regions. Therefore, the meshing density is larger in the cell model with smaller cell membrane thickness. Table 2 shows the number of meshing element in each specific region, including total suspension, cell membrane, intracellular fluid and extracellular fluid. In addition, a specific region surrounding the cell membrane was created to obtain the number of meshing elements around the cell membrane. It is obvious from Table 2 that the most meshing elements concentrate at the region around the cell membrane. The narrow region is reduced when the cell membrane thickness is enlarged, thus, the number of meshing elements is decreased when increasing cell membrane thickness.

Table 1. Meshing information for acceleration model.

	Original Model Meshing Information	Acceleration Model Meshing Information ($l_1 : l_2 = 1:1$)
Maximum element size (μm)	0.1	0.1
Minimum element size (μm)	0.01	0.01
Maximum element growth rate	1.20	1.20
Narrow factor	1	1

Table 2. The number of meshing element in different regions.

Regions	Original Model	Acceleration Model ($l_1 : l_2 = 1:1$)
Extracellular fluid	32,814	15,668
Intracellular fluid	30,566	14,748
Cell membrane	9712	4856
Region around cell membrane	70,528	33,620
Total suspension	73,092	35,272

3. Results and Discussion

3.1. D Single Spherical Cell Model

The full-mesh model cell membrane thickness is 5 nm. There are two simulations with equivalent thicker cell membrane (reduce-mesh model) thickness of 10 nm and 20 nm, respectively. The ratio between the full-mesh model cell membrane (cell membrane M_1 in Figure 3) thickness and the enlarged part of reduce-model cell membrane (cell membrane M_2 in Figure 3) thickness is 1:1 and 1:3, respectively. The simulation results are the progress of β-dispersion which reflects dispersions on relative permittivity and conductivity, as shown in Figures 4 and 5. The main characteristics of β-dispersion are the dispersion frequency range and the magnitude of the relative permittivity and conductivity [27,30,31]. As shown in Figure 4, the dispersion frequency is ranging from 100 kHz to 10 MHz. The magnitude of relative permittivity at low frequency and high frequency are 960 and 80, respectively. The magnitude of conductivity at low frequency and high frequencies were 0.925 S/m and 1.04 S/m, respectively. This work was aimed at accelerating the computing progress with high accuracy,

15 sample points are selected in Figure 4 to describe the main feature of an entire β -dispersion and verify the accuracy of the acceleration method. The results of acceleration models are validated by the full-mesh model to verify the accuracy of the acceleration method. The errors between the acceleration model and full-mesh model at each point in Figures 4 and 5 have been calculated. The calculated error and computing time for acceleration model and full-mesh model are shown in Table 3.

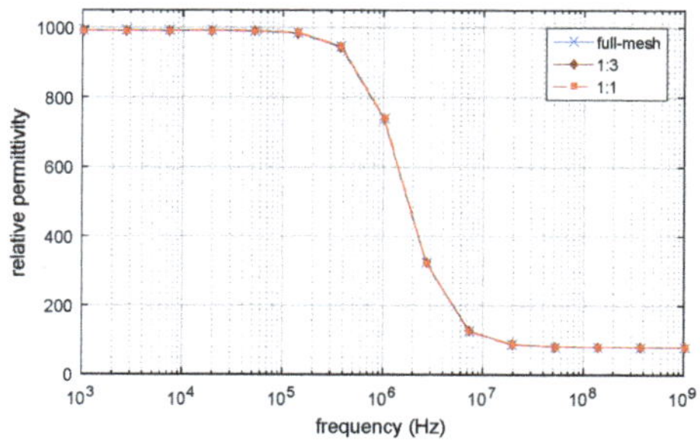

Figure 4. Permittivity of 2D single spherical cell model.

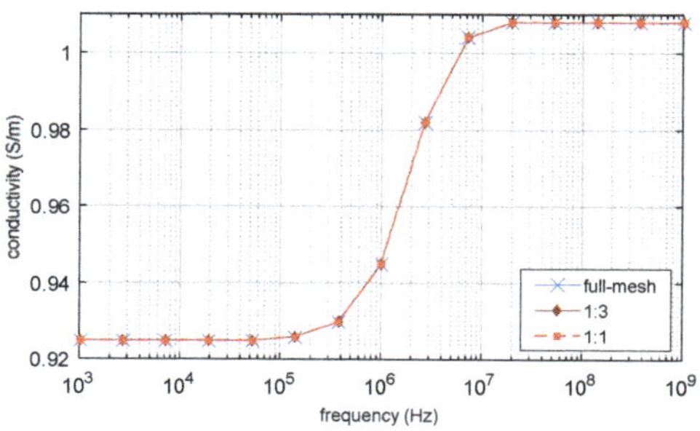

Figure 5. Of 2D single spherical cell model.

Table 3. Results of the 2D spherical cell model.

Thickness Ratio (M_1/M_2)	Number of Elements	Error	Computing Time (Minutes)
Original model	73,092	N/A	73
1:1	35,272	0.07%	28
1:3	17,342	0.2%	13

As shown Table 3, the number of elements of the full-mesh model can be reduced to 24% when the equivalent cell membrane (reduce-mesh model) is four times as the thickness of the full-mesh model cell membrane (M_1 cell membrane to M_2 cell membrane thickness ratio is 1:3). The computing time was reduced from 73 min to 13 min, with only a 0.2% error on the simulation result. The computing time of the reduce-mesh model is 18% of the full-mesh model. The spectroscopy of conductivity and relative permittivity is shown in Figures 4 and 5, the results of the reduce-model and full-mesh model show the same magnitude over the same frequency range with an error less than 0.2%. The β dispersion of reduce-model and full-mesh model both starts at 100 kHz and both ends at 10 MHz. This shows that the reduce-mesh model can significantly accelerate the computing progress with only a tiny error on the result in the 2D-FEM spherical model simulation.

3.2. D Cell Deformation Model

This simulation is to verify that the acceleration method not only works on spherical cell model but also works on the deformation model (oval model).

The deformation model is an oval model in order to simulation the cell deformation [28]. The electrical parameters of the oval model are the same as that of the custom spherical model. The model is shown in Figure 6, where a and b are the length of the semi-major and semi-minor axis, respectively. The length of the semi-major axis is set to be a = 12 μm and the length of the semi-minor axis is set to be b = 2 μm, k_a and k_c are the conductivity of the extracellular and intracellular fluid, respectively, ε_c and ε_a are the relative permittivity of the intracellular and extracellular fluid respectively, ε_m is the relative permittivity of cell membrane, d_m is the cell membrane thickness. The full-mesh model cell membrane thickness is still 5 nm and the electrical properties are calculated using equation 18. The simulation result of β dispersion is shown in Figure 7, Figure 8 and the parameters of the computing time and error are shown in Table 4.

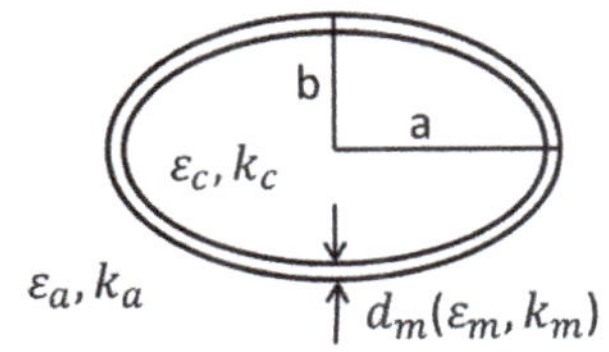

Figure 6. Deformation model.

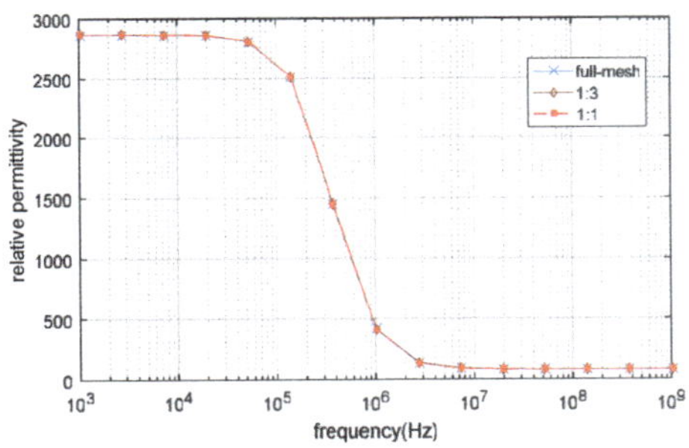

Figure 7. The relative permittivity of the 2D single deformation cell model.

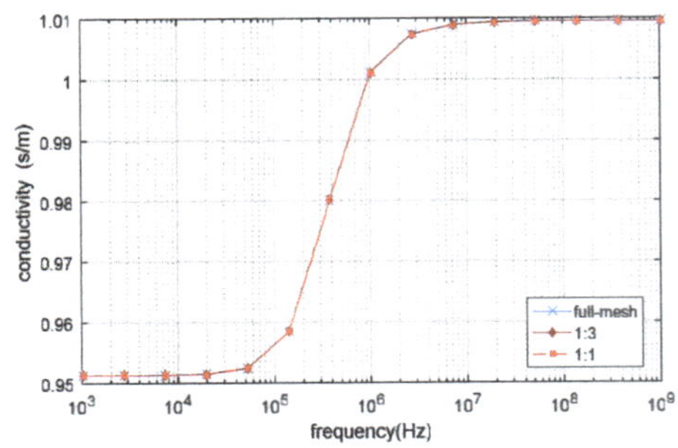

Figure 8. The conductivity of 2D single deformation cell model.

Table 4. Results of 2D Deformation Cell Model.

Thickness Ratio (M_1/M_2)	Number of Elements	Error	Computing Time
Original (a = 12, b = 2)	133,184	N/A	2 h 24 min
1:1	64,600	0.04%	62 min
1:3	32,097	0.11%	25 min

The number of elements of full-mesh model can be reduced to 24.1% as shown in Table 4 and the computing time is reduced from 2h 24mins to 25mins. The simulation error between reduce-mesh model and full-mesh model is only 0.11%. As shown in Figures 7 and 8, the results of the full-mesh model and reduce-mesh model exhibits β dispersion with tiny error in the same frequency range. The results show that the acceleration method is feasible to simulate not only the spherical model but also deformation model (oval model). The computing time is significantly reduced with acceptable error.

3.3. D Spherical Cell Model

The simulation model used for 3D FEM simulation is shown in Figure 2. 3-D single sphere cell model is used. The radius of the cell model is set to be 5 μm, and the cell membrane thickness is 5 nm. The top view of eddy current distributions at lower frequency 1 kHz and higher frequency 10 MHz are shown in Figure 9. The single-cell model blocks eddy currents at a lower frequency and becomes conductive at a higher frequency which meets the expectations of maxwell–wagner effect and other published works [9,11,15,29].

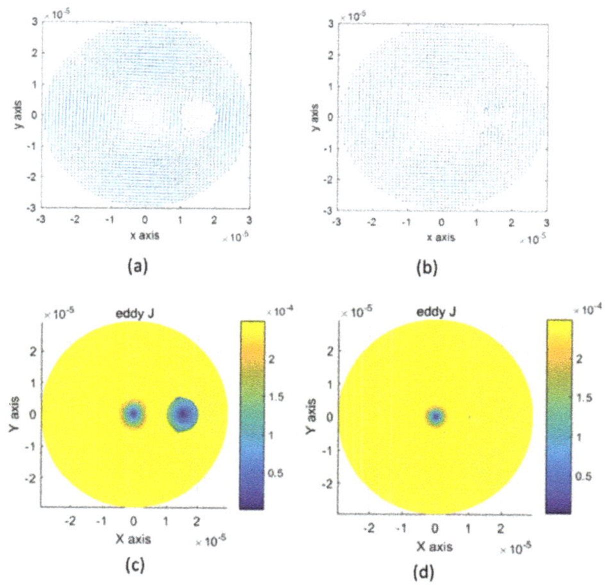

Figure 9. (**a**) Eddy current flow at low frequency 1 kHz. (**b**) Eddy current flow at high frequency 10 MHz. (**c**) Eddy current density at low frequency 1 kHz. (**d**) Eddy current density at high frequency 10 MHz.

The full-mesh model cell membrane thickness is 5 nm. There are two reduce-models with equivalent thicker cell membrane (reduce-mesh model) thickness of 10 nm and 20 nm, respectively. The ratio between the full-mesh model cell membrane thickness and the reduce-mesh cell membrane thickness is 1:1 and 1:3, respectively. The relative permittivity and conductivity result are shown in Figures 10 and 11. β dispersion can be observed from Figures 10 and 11. The β dispersion of the full-mesh model and reduce-mesh model has an error of 2% on the magnitude. The frequency range of the β dispersion of full-mesh model and reduce-mesh model are the same, ranging from 100 kHz to 10 MHz. The results, including the number of elements, the computing time and error of full-mesh model and reduce-mesh model, are shown in Table 5. The number of elements of the full-mesh model can be reduced to 28%, as shown in Table 5, and the computing is reduced from 4 h 37 min to 71 min. The simulation error between reduce-mesh model and full-mesh model is 2%. The error is larger than the two-dimensional simulation result, but it is acceptable for β-dispersion simulations. The results show that the acceleration method is feasible to simulate a three-dimensional spherical model. The computing time is significantly reduced with acceptable error.

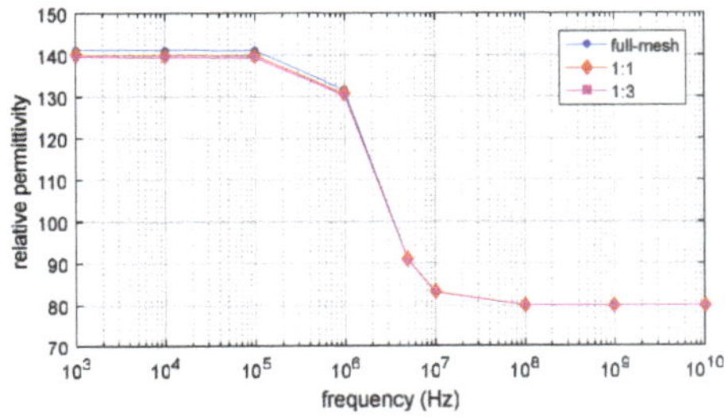

Figure 10. The relative permittivity of the 3D single spherical cell model.

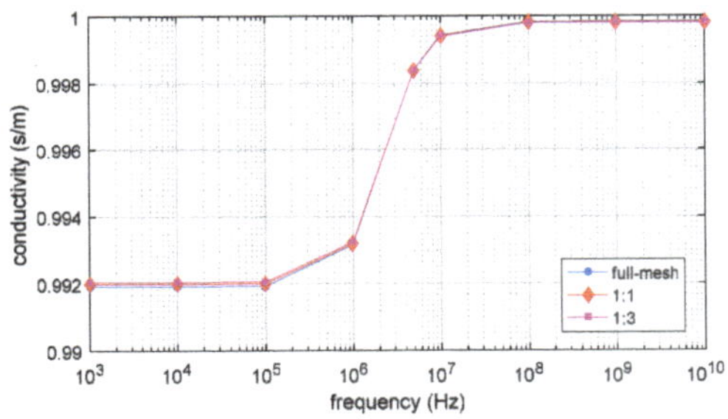

Figure 11. The conductivity of 3D single spherical cell model.

Table 5. Result of 3D Spherical Cell Model.

Thickness Ratio (M_1/M_2)	Number of Elements	Error	Computing Time
Original model	205,604	N/A	4 h 37 min
1:1	141,870	0.4%	3 h 5 min
1:3	57,934	2%	1 h 11 min

3.4. *Experimental Validation by Contact-Electrode Measurement*

The four-electrode measurement system is used to measure the impedance spectroscopy of biological samples. The electrode is contacted directly to the samples to measure the impedance over the samples using an impedance analyser. The measurement system work in the following way: the electrodes generate an input signal which applies an alternating electric field and current in the samples. By measuring the current and voltage across the samples, the conductivity and permittivity of the samples can be calculated. The impedance analyser used in this work is Solatron 1260, which is efficient and accurate over the frequency ranging from hundreds of hertz to 32 MHz. The effective area of the measurement electrodes is fixed to $S = 2\text{ cm} * 3\text{ cm} = 6\text{ cm}^2$, and the distance between the electric field measuring electrodes is fixed to L = 6.6 cm. The conductivity can then be calculated by $\sigma = 1/\rho = L/RS$. R is the real part of the measured impedance, ρ is the resistivity. The measurement sample is fresh potato (obtained from a local market). This measurement is to validate the FEM simulation result. All FEM simulation parameters are set to be the realistic value obtained from the measurement.

In Figure 12, the FEM and measurement results show similar β -dispersion with the same dispersion frequency range. There is some error on the magnitude of the conductivity between the measurement result and the simulation result. However, the error is acceptable for the simulation approach of the measurement result. The conductivity of measurement and simulation result is low at a lower frequency, and the dispersion starts at around 50 kHz. The dispersion ends at around 2 MHz and the conductivity of measurement and simulation result is increased to 0.15 S/m. The spectroscopy curve is flat over the higher frequency range. The measurement result validated the full-mesh and reduce-mesh FEM simulation result and agreed with other published works [9,13,15,21,25,32].

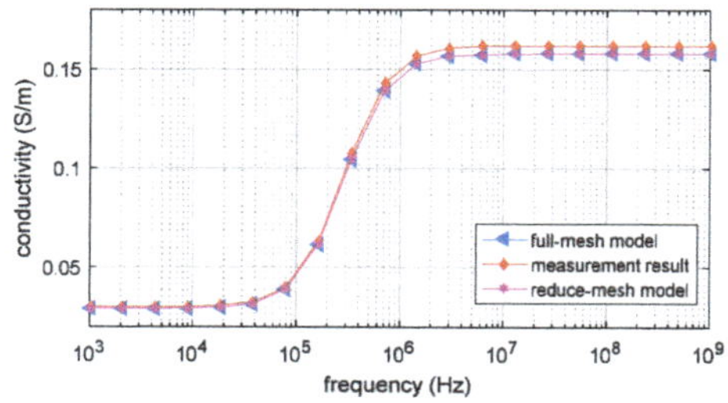

Figure 12. Impedance spectroscopy of FEM and measurement result.

3.5. Verification by Analytical Result

The analytical result can be calculated according to Maxwell–Wagner Equations (1) and (2). The analytical result is compared with the 3D finite element simulation result. So the factor v in equation (1) is defined as $v = \left(1 - \frac{d_m}{R}\right)^3$. According to Schwan [9], the impedance spectroscopy of the cell model can be described by Debye relaxation based on Equation (4). In Figures 13 and 14, both analytical and FEM results show the same magnitude and frequency range of beta dispersion with an error less than 0.1%. The volume fraction of the cell is set to be the same as P = 3.5%. The magnitude and frequency range of β-dispersion on the full-mesh model and reduce-mesh model agree with the analytical result. The result validates that the three-dimension FEM simulation on cell suspension is accurate and the improved acceleration method is also validated. The proposed acceleration method can be applied to further FEM simulations on irregular shape cell models and thin shell models.

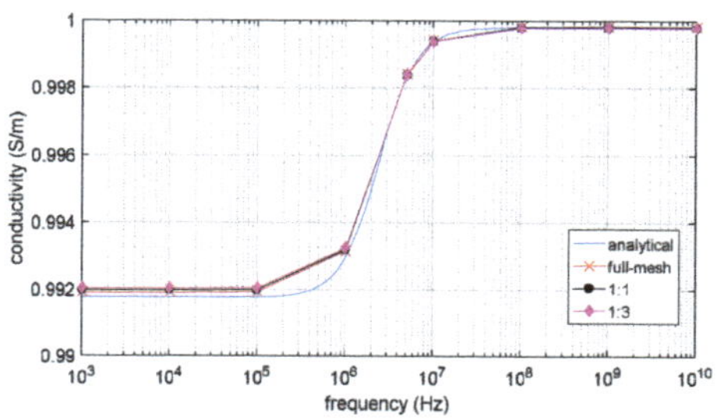

Figure 13. Calculated conductivity of the analytical solution and FEM simulation.

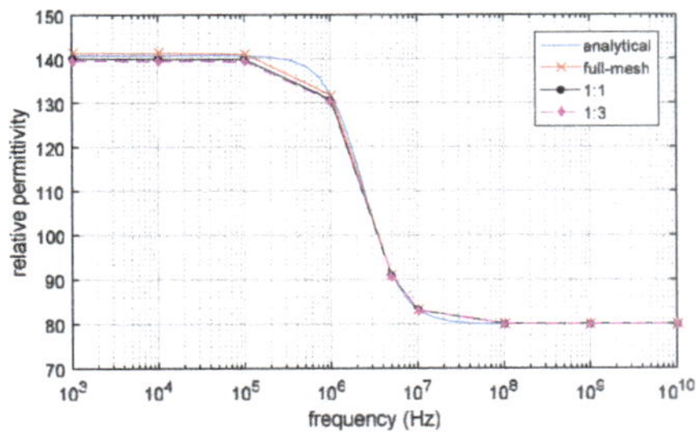

Figure 14. Calculated relative permittivity of the analytical solution and FEM simulation.

4. Conclusions

This paper proposed a method to accelerate FEM calculation with bio-cell models. The idea is to replace the thin cell membrane (full-mesh model) with an equivalent thicker cell membrane (reduce-mesh model). Then the number of meshing elements of the full-mesh model is reduced, and thus the computing time is reduced.

According to the simulation results, the reduce-mesh model can be used in both 2D and 3D FEM simulations. The amount of computing time is significantly reduced with an error no more than 0.5% in 2D simulation. The error on 3D simulation result is no more than 2%. All simulation results are validated by measurements and analytical results.

The proposed acceleration method is validated to be fast and accurate on cell models and the acceleration method has potential for all other thin shell FEM models.

Author Contributions: J.T. and W.Y. designed, planned and analysed the major acceleration method and prepared and reviewed the manuscript. M.L. helped in building the 3D FEM simulation server. Y.X. helped in building the experimental set-up for validation and reviewed the manuscript. All authors have read and agreed to the published version of the manuscript.

Funding: This research received no external funding.

Conflicts of Interest: The authors declare no conflict of interest.

References

1. Gimsa, J.; Schnelle, T.; Zechel, G.; Glaser, R. Dielectric spectroscopy of human erythrocytes: Investigations under the influence of nystatin. *Biophys. J.* **1994**, *66*, 1244–1253. [CrossRef]
2. Awayda, M.S.; Driessche, W.; Helman, S.I. Frequency-dependent capacitance of the apical membrane of frog skin: Dielectric relaxation processes. *Biophys. J.* **1999**, *76*, 219–232. [CrossRef]
3. Mukhopadhyay, S.C.; Gooneratne, C.P. A novel planar-type biosensor for noninvasive meat inspection. *IEEE Sens. J.* **2007**, *7*, 1340–1346. [CrossRef]
4. Jaffrin, M.Y.; Morel, H. Body fluid volumes measurements by impedance: A review of bioimpedance spectroscopy (BIS) and bioimpedance analysis (BIA) methods. *Med. Eng. Phys.* **2008**, *30*, 1257–1269. [CrossRef] [PubMed]
5. Morimoto, T.; Kimura, S.; Konishi, Y.; Komaki, K.; Uyama, T.; Monden, Y.; Kinouchi, D.Y.; Iritani, D.T. A study of the electrical bio-impedance of tumors. *J. Investig. Surg.* **1993**, *6*, 25–32. [CrossRef] [PubMed]
6. Kerner, T.E.; Paulsen, K.D.; Hartov, A.; Soho, S.K.; Poplack, S.P. Electrical impedance spectroscopy of the breast: Clinical imaging results in 26 subjects. *IEEE Trans. Med. Imaging* **2002**, *21*, 638–645. [CrossRef] [PubMed]
7. Romsauerova, A.; McEwan, A.; Horesh, L.; Yerworth, R.; Bayford, R.H.; Holder, D.S. Multi-frequency electrical impedance tomography (EIT) of the adult human head: Initial findings in brain tumours, arteriovenous malformations and chronic stroke, development of an analysis method and calibration. *Physiol. Meas.* **2006**, *27*, S147. [CrossRef]

8. Schwan, A.; Herman, P. Electrical properties of tissue and cell suspensions. *Adv. Biol. Med. Phys.* **1957**, *5*, 147–209.
9. Schwan, A.; Herman, P. Electrical properties of tissues and cell suspensions: Mechanisms and models. In Proceedings of the 16th Annual International Conference of the IEEE Engineering in Medicine and Biology Society, Baltimore, MD, USA, 3–6 November 1994.
10. Damez, J.L.; Clerjon, S.; Abouelkaram, S.; Lepetit, J. Dielectric behavior of beef meat in the 1–1500 kHz range: Simulation with the Fricke/Cole–Cole model. *Meat Sci.* **2007**, *77*, 512–519. [CrossRef]
11. Stubbe, M.; Gimsa, J. Maxwell's Mixing Equation Revisited: Characteristic Impedance Equations for Ellipsoidal Cells. *Biophys. J.* **2015**, *109*, 194–208. [CrossRef]
12. Asami, K.; Tetsuya, H.; Naokazu, K. Dielectric approach to suspensions of ellipsoidal particles covered with a shell in particular reference to biological cells. *Jpn. J. Appl. Phys.* **1980**, *19*, 359. [CrossRef]
13. Pauly, H.; Schwan, H.P. Impedance of a suspension of ball-shaped particles with a shell; a model for the dielectric behavior of cell suspensions and protein solutions. *Z. Naturforsch.* **1959**, *14*, 125. [CrossRef]
14. Tuncer, E.; Stanisław, M.; Gubański, C.; Nettelblad, B. Dielectric relaxation in dielectric mixtures: Application of the finite element method and its comparison with dielectric mixture formulas. *J. Appl. Phys.* **2001**, *89*, 8092–8100. [CrossRef]
15. Asami, K. Dielectric dispersion in biological cells of complex geometry simulated by the three-dimensional finite difference method. *J. Phys. D Appl. Phys.* **2006**, *39*, 492. [CrossRef]
16. Mejdoubi, A.; Brosseau, C. Finite-element simulation of the depolarization factor of arbitrarily shaped inclusions. *Phys. Rev. E* **2006**, *74*, 031405. [CrossRef]
17. Takashima, S.; Asami, K.; Takahashi, Y. Frequency domain studies of impedance characteristics of biological cells using micropipet technique. *Biophys. J.* **1988**, *54*, 995. [CrossRef]
18. Fricke, H. The Maxwell-Wagner dispersion in a suspension of ellipsoids. *J. Phys. Chem.* **1953**, *57*, 934–937. [CrossRef]
19. Jin, J.M. *The Finite Element Method in Electromagnetics*; Wiley-IEEE Press; John Wiley & Sons: Hoboken, NJ, USA, 2015; pp. 1–876.
20. Bíró, O. Edge element formulations of eddy current problems. *Comput. Methods Appl. Mech. Eng.* **1999**, *169*, 391–405. [CrossRef]
21. O'Toole, M.D.; Marsh, L.A.; Davidson, J.L.; Tan, Y.M.; Armitage, D.W.; Peyton, A.J. Non-contact multi-frequency magnetic induction spectroscopy system for industrial-scale bio-impedance measurement. *Meas. Sci. Tech.* **2015**, *26*, 035102. [CrossRef]
22. Georgii, J.; Rudiger, W. A streaming approach for sparse matrix products and its application in Galerkin multigrid methods. *Electron. Trans. Numer. Anal.* **2010**, *37*, 263–275.
23. Fritschy, J.; Horesh, L.; Holder, D.S.; Bayford, R.H. Using the GRID to improve the computation speed of electrical impedance tomography (EIT) reconstruction algorithms. *Physiol. Meas.* **2005**, *26*, S209. [CrossRef]
24. Lu, M.Y.; Anthony, P.; Yin, W.L. Acceleration of Frequency Sweeping in Eddy-Current Computation. *IEEE Trans. Magn.* **2017**, *53*, 1–8. [CrossRef]
25. Gheorghiu, E. Measuring living cells using dielectric spectroscopy. *Bioelectrochem. Bioenerg.* **1996**, *40*, 133–139. [CrossRef]
26. Sekine, K. Application of boundary element method to calculation of the complex permittivity of suspensions of cells in shape of D∞ h symmetry. *Bioelectrochemistry* **2000**, *52*, 1–7. [CrossRef]
27. Beving, H.; Eriksson, L.E.G.; Davey, C.L.; Kell, D.B. Dielectric properties of human blood and erythrocytes at radio frequencies (0.2–10 MHz); dependence on cell volume fraction and medium composition. *Eur. Biophys. J.* **1994**, *23*, 207–215. [CrossRef] [PubMed]
28. Di Biasio, A.; Cametti, C. Effect of shape on the dielectric properties of biological cell suspensions. *Bioelectrochemistry* **2007**, *71*, 149–156. [CrossRef]
29. Saville, D.A.; Bellini, T.; Degiorgio, V.; Mantegazza, F. An extended Maxwell–Wagner theory for the electric birefringence of charged colloids. *J. Chem. Phys.* **2000**, *113*, 6974–6983. [CrossRef]

30. Davis, L.C. Polarization forces and conductivity effects in electrorheological fluids. *J. Appl. Phys.* **1992**, *72*, 1334–1340. [CrossRef]
31. Zhou, Y.Y.; Wang, A.P.; Zhou, P.; Wang, H.; Chai, T.Y. Dynamic performance enhancement for nonlinear stochastic systems using RBF driven nonlinear compensation with extended Kalman filter. *Automatica* **2020**, *112*, 108693. [CrossRef]
32. Tang, J.W.; Yin, W.L.; Lu, M.Y. Bio-impedance spectroscopy for frozen-thaw of bio-samples: Non-contact inductive measurement and finite element (FE) based cell modelling. *J. Food Eng.* **2020**, *272*, 109784. [CrossRef]

biosensors

MDPI

Article

Development of an Optical Method for the Evaluation of Whole Blood Coagulation

Marinos Louka and Efstathios Kaliviotis *

Department of Mechanical Engineering and Materials Science and Engineering, Cyprus University of Technology, 3036 Limassol, Cyprus; ma.louka@edu.cut.ac.cy
* Correspondence: e.kaliviotis@cut.ac.cy

Abstract: Blood coagulation is a defense mechanism, which is activated in case of blood loss, due to vessel damage, or other injury. Pathological cases arise from malfunctions of the blood coagulation mechanism, and rapid growth of clots results in partially or even fully blocked blood vessel. The aim of this work is to characterize blood coagulation, by analyzing the time-dependent structural properties of whole blood, using an inexpensive design and robust processing approaches. The methods used in this work include brightfield microscopy and image processing techniques, applied on finger-prick blood samples. The blood samples were produced and directly utilized in custom-made glass microchannels. Color images were captured via a microscopy-camera setup for a period of 35 min, utilizing three different magnifications. Statistical information was extracted directly from the color components and the binary conversions of the images. The main advantage in the current work lies on a Boolean classification approach utilized on the binary data, which enabled to identify the interchange between specific structural elements of blood, namely the red blood cells, the plasma and the clotted regions, as a result of the clotting process. Coagulation indices produced included a bulk coagulation index, a plasma-reduction based index and a clot formation index. The results produced with the inexpensive design and the low computational complexity in the current approach, show good agreement with the literature, and a great potential for a robust characterization of blood coagulation.

Keywords: blood coagulation; image sensing; image classification

Citation: Louka, M.; Kaliviotis, E. Development of an Optical Method for the Evaluation of Whole Blood Coagulation. *Biosensors* **2021**, *11*, 113. https://doi.org/10.3390/bios11040113

Received: 2 March 2021
Accepted: 6 April 2021
Published: 9 April 2021

1. Introduction

Blood flow allows the distribution of oxygen and nutrients to the organs and tissues. The importance of blood circulation is obvious in cases where health related problems affect people's daily life. For instance, the conditions of hemophilia and hyper-coagulation have a significant impact on patients mortality rate. Hemophilia can cause major internal and external uncontrollable bleeding from minor injuries, due to the inability of proper clot formation. On the other hand, hyper-coagulation is associated with abnormal and uncontrollable formation of blood clots in the absence of injuries, or the inability of fibrinolysis to control clot growth and clot dissolution. Both conditions cause pathological problems and patients in many cases need daily monitoring in order to regulate their medication [1].

Hemostasis is a physiological process in which blood loss due to vessel injury is minimized by clotting, so that blood flow is maintained normal in the vessels. The hemostasis process involves the mechanisms of clot formation at the site of the injury, the prevention of uncontrollable clot growth, and the lysis (dissolution) of the clot. This process is characterized by a series of reactions which involve platelets, red blood cells (RBCs), coagulation factors and anticoagulants in a three-stage procedure. In the first stage of the process, named the primary hemostasis, an initial soluble white clot is created, consisting mainly of activated platelets. Activated platelets, at the site of the endothelial injury, bind with Von Willebrand factor through the GPIb-receptor [2]. Von Willebrand is a glycoprotein produced in the endothelial cells and it is critical to the initial clotting as it helps the aggre gation of the platelets at the site of the injury. In the next stage, the secondary hemostasi (also known as blood coagulation), the inactive coagulation factors, that exist in the plasm under physiological conditions, are activated. In a complex process the coagulation factor promote the transformation/conversion of the soluble white clot and fibrinogen into a insoluble, stable clot and fibrin mesh, that plugs the damaged vessel [3,4].

Blood coagulation is described by the coagulation Cascade (or Waterfall model which includes the intrinsic and extrinsic pathways, both leading to the common pathwa The final stage of hemostasis regards the deactivation of the coagulation factors and th dissolution of the clot to prevent oversize and unnecessary clotting at the site of th injury, which can affect the blood flow. This stage is known as fibrinolysis and consist o natural inhibitors that terminate the growth of clots and break them down. All three of th aforementioned stages require proteins, coagulation factors and natural anticoagulants t initiate and complete hemostasis [5]. Imbalance between coagulation and anticoagulatio proteins as well as deficiencies in any of them, can cause bleeding disorders like hemophili A, hemophilia B and Von Willenbrand disease, or arterial/venous thrombosis [6–8].

When blood is removed from the body and placed in a glass test tube, red bloo cells (RBCs) are able to aggregate due to low shear stress, blood clots fairly quickly, an the blood coagulation process can be assessed immediately. Red blood cell aggregatio (RBCA) causes the formation of stacks of RBCs in coin-like structures (termed rouleaux the formation of larger aggregates from rouleaux combination, and even a 3-dimensiona network of aggregates [9]. The result of the contraction of RBCs to aggregates is th appearance of larger plasma regions in the blood sample. Calcium ions are required for th clotting process, whereas acidic substances like EDTA (ethylenediaminetetraacetic acid) o citrate can prevent the formation of clots by binding calcium. EDTA and citrate are use widely in the Hematology laboratory as anticoagulants. Clotting can be initiated in vitro a a later time by adding back an excess of calcium ions [10].

A variety of laboratory equipment and point-of-care (POC) devices have been de veloped through the years for blood coagulation assessment, exploiting a range of bloo properties [11]. In those tests, platelet count, functionality, adhesion and aggregatio are of primary importance. Platelet testing provides valuable information concernin patients with bleeding disorders; it allows to monitor patient's medication treatment, as sess in-surgery hemostasis and guide transfusion medicine with manual or automate instruments [12]. The cost of hand-held POC devices and their disposable strips is no negligible, whereas time and complexity might be an issue for the techniques (such a thromboelastography) that exploit the mechanical properties of blood coagulation.

The extrinsic and intrinsic pathways can be used for in vitro laboratory test screening extracting parameters like the prothrombin time (PT) and activated partial thromboplasti time (aPTT), respectively [13]. PT and aPPT indicate the time that it takes for clots to form in plasma when calcium and tissue thromboplastin are present, and their magnitude depend on device specific parameters. The extrinsic pathway test, aPTT, is sensitive t factors, I, II, V, VII, X, whereas the intrinsic pathway test, PT, is sensitive to factors XII, XI IX, VIII, and further, sensitive to deficiencies of prekallikrein and high molecular weigh kininogen (HMWK). These indices can be derived with a variety of instruments, with each of the instrument exploiting different blood coagulation properties [14]. Also, an important widely used index for blood coagulation is the International Normalize Rati (*INR*) which is derived from the PT test and it is used for effective monitoring patient under oral anticoagulation therapy [15]. *INR* is calculated by the formula:

$$INR = \left(\frac{PT_{patient}}{MNPT_{patient}}\right)^{ISI} \quad (1)$$

where $MNPT_{patient}$ is the mean normal PT of a control group and ISI is the International Sensitivity Index, which is used to adjust sensitivity difference of a variety of reagents used by instruments.

Electrochemical techniques for analysing coagulation are widely adopted in Hematology laboratories and POC devices. These techniques are based on resistance/capacitance changes of whole blood as clot related macromolecules react to the added agents in the sample. During the coagulation process the dominant impedance is moved from the plasma impedance to the aggregated red blood cells, and finally to the fibrin network impedance [16]. As blood coagulation escalates in time, the total impedance increases and can be accurately measured and provide the *INR* index, using the formula in Equation (1). This method leads to smaller and patient friendly POC devices as it is based on integrated electronics platforms and disposable test strips, with examples like CoaguChek$^{(r)}$ XS (Roche Diagnostics, Indianapolis, IN, USA), microINR (iLine$^{(r)}$ Microsystems S.L., San Sebastian, Spain) and i-STAT (Abbott Point of Care Inc., Abbott Park, IL, USA). However as mentioned, the cost of such devices and their disposable strips is not negligible [17,18].

Another approach to assess further details of the blood coagulation process is the Thromboelastography method ($TEG^{(r)}$, Haemonetics, Braintree, MA, USA), which examines the mechanical characteristics of blood in terms of viscoelasticity and kinetics as blood clots form. TEG was first introduced by Hertert in 1948 [19], as a global essay that characterizes in real time the creation of clots in whole blood using a container for the sample and a constant forced rotating rod immersed into the sample. The rod is connected to a transducer, which measures the torque generated by the developed clot. Rotational thromboelastometry ($ROTEM^{(r)}$ TEM International, Munich, Germany) is considered the evolution of TEG, following the same fundamental principles. Both exploit the viscoelasticity on whole blood as coagulation and lysis occurs with the use of reagents, producing essentially similar indices based on characteristics of the generated curves. The basic parameters in TEG and ROTEM need 30 to 60 min and are calculated in terms of the time-evolved curve amplitude (in mm) which indicates the clot firmness. Coagulation indices are derived from linear combination of a variety of the parameters like the maximum reached amplitude, kinetic time, clotting initiation time and the angle between the initiation of coagulation tangent to the developed curve [20]. It is also suggested that Thromboelastography can potentially help to guide patients needs for transfusions in surgery and trauma induced coagulopathies, through specialized algorithms embedded in the devices [21,22].

Coagulation in blood causes intense structural changes that can be detected by image analysis techniques. Machine vision approaches offer a range of methods and techniques for image processing [23], and are used for the development of Lap-on-Chip devices [24]. These techniques can be used for real-time monitoring and characterization, in biomedical research and clinical diagnostics [24]. Methods such as image segmentation allow the identification of structures with specific characteristics [25,26]. To accommodate the blood sample to be processed, simple microfluidic configurations are often used, offering several advantages [27–30]. Other techniques, such piezocoagulography, have been also utilized for blood coagulation, showing also the possibility of studying the initial stage of fibrin formation in the clotting process [31].

In recent works presented in the literature, the optical properties of coagulating blood are exploited, by measuring the light propagation and/or scattering through a sample (please see [11] for a review on the subject). The produced time-varying curves provide valuable information about the blood coagulation process. This technique, termed photometry, requires the acquisition of sequential light patterns, usually images, utilising a light source and a camera. Photometric results can be achieved with monochromatic or polychromatic light sources, and experimental setups covering the infrared to the x-ray spectrum, utilizing appropriate cameras. It has been shown that light sources with a wavelength around 690 nm (near-infrared window) provide very good results, due to the minimal light absorption by hemoglobin and water [32]. During coagulation, the light patterns acquired by the camera change dynamically as clots develop in the sample.

Furthermore, light transmission through the sample is blocked by the newly formed clots and the averaged detected light is decrease due to the interaction between the platelets, the RBCs and the procoagulant plasma proteins [33,34]. The aforementioned phenomena can be characterized, after digitally process the captured images. Several studies in the literature utilize laser speckle analysis between consecutive images, and/or between a reference image (usually the initial one), in combination with cross- or auto-correlation techniques, which can be computationally demanding. The results are presented as time-varying curves with the potential of dividing them into different coagulation phases [35–37]. Indices extracted from those curves can describe blood coagulation and typically are validated by comparison with commercially available coagulometers [38,39]. Techniques, such us the aforementioned, utilize color spectra to assess blood coagulation and extract useful indices, by using a specific light wavelength. The produced information can be visualized by using pseudo-colored maps and as a result, they may lack a direct representation of the whole blood clotting process. Moreover, such techniques require specialized components and set-ups, such as laser diodes, optical components (splitters, expanders, polarizers, etc.).

In this work, an optical assessment technique for blood coagulation is presented, using whole blood from a healthy volunteer, with conventional bright-field microscopy. The tests were conducted using capillary whole blood obtained by finger prick without added reagents. Image processing algorithms have been developed to analyze a series of images acquired from a camera/microscope setup, in gray, color and binary levels, using three magnification lenses and custom-made microchannels. The produced curves display three distinct phases, the RBCA, the clot development and the completion of whole blood coagulation. A classification algorithm separates the images, by detecting the areas in which clots are developed. Results extracted from the four color channels (red, green, blue and gray scale) show good sensitivity and conclude that this method can be used to assess whole blood coagulation with low computational demands and inexpensive equipment. The current work is set in order to demonstrate the validity and the potential of the proposed approach.

2. Materials and Methods

2.1. Sample Preparation

The blood donation protocol for this work was approved by the Cyprus National Bioethics Committee (permission ref.: EEBK/EP/2019/19). A drop of blood was collected through a finger prick from a healthy volunteer (following 10 h of fasting) after the application of intermittent pressure to a finger in order to increase the blood flow. Then the area was cleaned with alcohol and a sterilized lancet was used to puncture the side of the fingertip. The first drop of blood was wiped with a sterilize cotton ball and the second drop of blood was placed at the entrance of a custom-made microchannel. The blood drop was then drawn by capillary forces in the observation area. A total of three different experiments were made and presented in this work, at three different microscope objectives. Each set of tests was completed in approximately two hours.

2.2. Experimental Setup

The microchannels were composed by two microscope glass slides (75 mm × 25 mm × 1 mm, for length, width and height, respectively) and double-faced adhesive tapes (Tesa$^{(r)}$) with dimensions 12.7 mm width and approximately 100 µm thickness. The tape was adhered in pairs and at a small angle between them, on one slide (Figure 1), creating an open space. The second glass slide was placed above the first one and adhered on the tapes, creating the microchannel and effectively the testing site. Two testing sites were created on the same glass slide. The glass slides overlapped for approximately 5 mm length, with entrance and exit widths of approximately 5 and 3 mm respectively, resulting in a small microchannel volume of approximately 2 mm^3. The angle of the tapes and the relatively small distance from the entrance to the exit resulted in fast filling, and fast deceleration and immobilization of the blood in the channel. After the placement of the sample in the channel, sunflower oil was applied at the entrance of the microchannel to isolate the sample from air, and restrict undesirable effects, such as air bubble formation and blood movement. The viewing window was at the center of the geometry at distances of approximately 2 mm from the entrance and exit. This region of interest in the center of the geometry appeared to have uniform characteristics, not influenced by the boundaries of the channel.

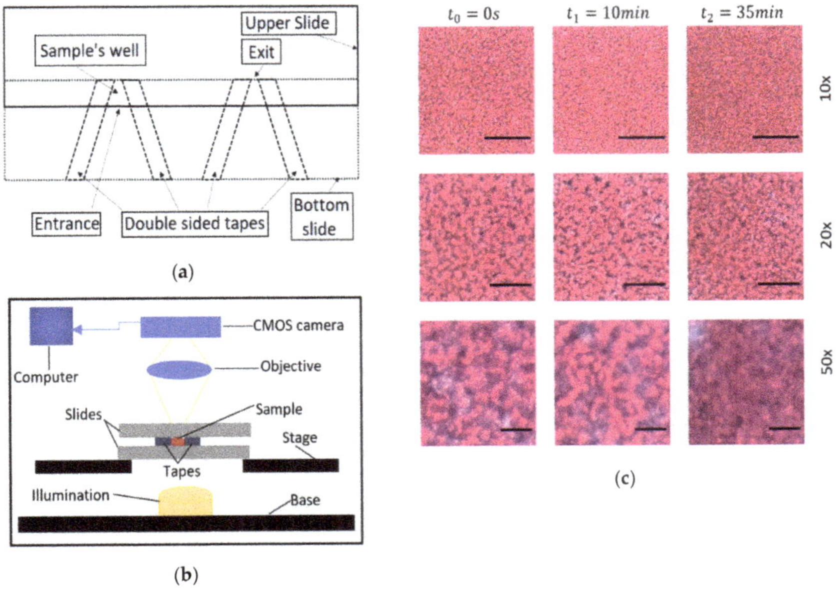

Figure 1. (**a**) The design of the microchannels. (**b**) Schematic of the experimental setup, which is based on forward propagation of light into to the sample, through the objective and into a CMOS camera for the acquisition of data. (**c**) Representative images showing the initial, an intermediate and a final state of coagulated blood at the 10×, 20× and 50× magnification objectives. Scale bars equal to 200, 100 and 40 μm for the 10× 20× and 50× objectives, respectively.

For video/image acquisition a JVC TK-C1380 camera was used, integrated on an Olympus BX51 microscope (Figure 1). The microscope light illumination (100 W long-life halogen bulb) was controlled by an intensity knob which was calibrated and scaled utilising an adhered protractor around the knob. Tests were conducted to validate the linearity between the detected light intensity and the protractor's degrees, with a deviation of around ±5 degrees. So, the protractor was utilized to set the light intensity appropriately, with a fine approximation in degrees. Before sampling, the camera focus was optimized manually by adjusting the microscope stage, and by observing the camera image to avoid opaqueness and loss of detail.

The camera was connected to a computer in which the video was saved. Three experiments were conducted with different magnifications lenses of 10×, 20× and 50× of numerical apertures of 0.25, 0.40 and 0.50, respectively, using blood sample from the same volunteer. Each acquired video has a 35 min duration using the free software VirtualDub at 5 frames per second (fps), enough to detect the blood coagulation process. Representative images from the three magnifications and at different times are shown in Figure 1. The resolution of the camera/microscopy setup was 1.0 μm/pixel, 0.55 μm /pixel and 0.22 μm/pixel ±10% for the 10×, 20× and 50× respectively, the observation image area was 576 × 768 pixels, and the processed image area were 500 × 500 pixels.

2.3. Whole Blood Sample Coagulation Evaluation

Blood coagulation was evaluated by digitally processing the acquired videos utilizin the developed algorithms (The MathWorks Inc., Matlab-Natick, MA, USA). The vide acquisition rate was initially set to 5 fps, however various reduced sampling rates were use in the processing stage in order to minimize storage needs and optimize the performance i terms of overall processing time. Apart from various sampling rates, different image size were used for performance optimization in the processing stage, while keeping the mai body of the algorithm unchanged. After the evaluation of the produced time-depende indices, the minimum frame rate, preserving the details of the process, was found to be at 4 per frame (0.25 fps), with total number of images per experiment to be 525 colored image

The processing stage can be divided into three main sections, with each one calculatin various values and parameters of interest, from the time-varying image series. First, eac image is separated into their red, green, blue and gray-scale (RGBGs) color channels. Fo each of the four color channels the mean intensity (MI) value is calculated for each imag and for all sequential images from the various tests. The MI corresponds to the average ligh travel through the sample at each specific moment. MI curves for all color channels wer smoothed in a moving average manner in order to extract more meaningful information a the initial parts of the process (up to approximately 10 s), where the data appear noisie The data were further normalized with the maximum values of the time series (and denote as MI^*), so that their time response of the various cases can be directly comparable.

In the second algorithmic section, the images were converted into binary data (i.e black and white pixels), using a unique threshold for each color channel, and for eac experiment. In the initial stages of the process, when no clots are present, the binary image have pixels with values one (white pixel) and zero (black pixel), representing the plasm and RBCs areas, respectively. This is due to the fact that the transmission of light throug the plasma is higher than the transmission of light through the RBCs and clots (see fo example the sample image at t_0, and at 50× magnification in Figure 1). So, as RBCs bloc more light than the plasma, their gray scale values is lower to that of the plasma pixel In the transformation of a gray scale image into binary the threshold value is the mos important parameter. The threshold is a gray scale value that separates all the pixels int two categories, those above and those below the threshold value. The RGBG threshold were based on the image characteristics in the peak value of their MI^* curve. Tha threshold value was selected for two reasons. First, at that point in time the phenomeno of RBCA is mostly complete, as it takes around three-four minutes to fully develop, an blood coagulation is the main function that affects the structure of the blood sample. Unti this point an increase in the averaged light sensing was observed. Thus, this time point ca be referred to as the clotting start (CS) time after which blood coagulation can be studie Secondly, by deriving the binarization threshold value from the CS correspond imag (resulting in a high threshold value), ensures the efficient detection of the growing clots i the subsequent analysis, as the distribution of pixel intensity will lean towards lower value due to blood coagulation. As the new structures, the clots, are forming they eventuall assigned zero pixel values, due to their light-blocking (absorbing) nature. Thus, the are occupied by clots increase in size, by the action of coagulation factors that blood plasm contains, and the corresponding plasma area is decreased. As a result, the percentage o the white pixels correspond to the plasma area is decreased in time. Therefore, the black pixels with zero values after processing, represent RBCs (and their aggregates) and the newly formed clots. A significant advantage of the present algorithm is that RBCs and newly formed clot areas can be distinguished and properly analyzed.

In the third section of the algorithm a Boolean algebra calculation loop was developed to extract indices from the binary image series with reference to the image at the CS time, in a pixel by pixel manner, similarly to methods used in studying elastic optical networks [40] Karnaugh mapping (KM) was used to simplify the expressions using logical operators. As mentioned above, the clot detection was classified as the change of pixels with values one to zero (i.e., from plasma pixel to clots pixel). As each binary image only consists of two values, ones and zeros, there are four possible combinations when comparing two images, each corresponding to a specific condition/stage regarding the blood coagulation process. The first combination is the zero to zero pixel value that regards the detection of RBCs or RBCA region that remains unchanged. The second regards changes from a zero value to one, denoting a new plasma area (NPA) which can appear due to the RBC aggregation (RBCA), in which RBCs attract each other leaving an empty from cells space (plasma). This procedure however is considered complete after the initial 4–5 min as mentioned above. In a third combination, a change from one to zero corresponds to the clot formation (CF) in areas where previously were occupied by plasma. In the fourth combination, values compared from one to one represent the plasma area (PA) still unoccupied by blood clots. As in the above section of the algorithm, the change of a white pixel to a black pixel is to be expected and indicates the activity of the coagulation process in the sample. For the four pixel comparisons mentioned above, the image corresponding to CS time is used as the reference image. Using the results of the pixel comparison the algorithm can separate areas as RBCs, the newly formed clots (CF), the initial plasma area (PA) at CS, and the new plasma areas (NPA).

Table 1 shows the truth table of the aforementioned combinations, and gives information about the four outputs, which can be conceived as four new images. Each new image has binary pixel values according to the output.

Table 1. The four combinations of the pixel-wise comparison between the image I at reference time τ (denoted as $I(\tau)_{i,j}$, and the subsequent images $I(\tau+t)_{i,j}$.

Inputs		**Outputs**			
$I(\tau)_{i,j}$	$I(\tau+t)_{i,j}$	$RBCs_{i,j}$	$NPA_{i,j}$	$CF_{i,j}$	$PA_{i,j}$
0	0	1	0	0	0
0	1	0	1	0	0
1	0	0	0	1	0
1	1	0	0	0	1

$I(\tau)_{i,j}$ denotes the value of the pixel in the *i*th row, *j*th column of the reference image and $I(\tau+t)_{i,j}$ the corresponding pixel of the sequential image at a time $\tau+t$, where τ is the CS time. The outputs RBCs, NPA, CF and PA are new binary data (images) at a given time, after CS, and they have the same size as the compared images, representing specific structural conditions in blood. The information from the Table 1 and the utilization of KM are used to derive indices describing, and quantifying, each of the aforementioned conditions in every image. The indices are developed according to Equations (2)–(5), with N noting the number of pixels of the image (total number of columns N_i and total number of rows N_j are equal to N, $\bar{I}$ denoting the complement pixel value of I at that position (logical NOT operation), and with the symbol ^ denoting the logical AND operation. For each combination and for all images, the total number of ones is divided by the total number of pixels in the image, resulting in the expression of the indices as fractional quantities (denoted with the * superscript).

$$RBCs^{*} = \frac{1}{N}\sum_{i=1}^{N}\sum_{j=1}^{N}\bar{I}(\tau)_{i,j}\hat{\bar{I}}(\tau+t)_{i,j} \quad (2)$$

$$NPA^{*} = \frac{1}{N}\sum_{i=1}^{N}\sum_{j=1}^{N}\bar{I}(\tau)_{i,j}\hat{I}(\tau+t)_{i,j} \quad (3)$$

$$CF^{*} = \frac{1}{N}\sum_{i=1}^{N}\sum_{j=1}^{N}I(\tau)_{i,j}\hat{\bar{I}}(\tau+t)_{i,j} \quad (4)$$

$$PA^* = \frac{1}{N}\sum_{i=1}^{N}\sum_{j=1}^{N} I(\tau)_{i,j}\hat{I}(\tau+t)_{i,j} \quad (5)$$

The addition of Equations (3) and (5) represents the total apparent blood plasma at any given time, after the CS time. Moreover, one can see that the pairs Equations (2) and (5), and Equations (3) and (4) are complementary due to their definition from Table 1. Therefore, it is of interest to further analyse the relation between $RBCs^*$ and CF^*. Figure 2a shows sections from a reference and a compared image, and the four output matrices. Figure 2b illustrates the behavior of $RBCs^*$ and CF^*. As mentioned earlier, the RBCA process was mostly complete at CS, as normally reaches a steady state after around 120 s. However, this time period could be extended due to various reasons. For example, small cell movements inside the microchannel could contribute to a prolonged RBCA time. After CS, a small activity of RBCA was observed for a small period of time, and that resulted in a small increase in NPA^* and a small decrease in CF^*. The apparent increase of $RBCs^*$ after it has reached a minimum (which denotes the completion of RBCA) seems artificial, since it is caused by the re-appearance of a zero value pixel area that should correspond to CF^*, at the same location where previous NPA^* was detected. In terms of pixel values, a zero in the reference image had initially compared with one (corresponding pixel in the subsequent image) but at a later time it was compared with a zero again. However, this condition most probably corresponds to the detection of clots in the region of NPA and not to a new RBCs. In a correction process, the increase that occur from the re-appearance of $RBCs^*$ is subtracted, and the subsequent increase is added to CF^* as Figure 2b shows.

In order to distinguish the most efficient image resolution and color channel for the image processing, a sensitivity index (SI) was developed (Equation (6)) to compare the resulted curves. The SI was applied only to the plasma reduction (PR) and the CF curves, as the MI^* curves are normalized using their own limits. PR is defined as the ratio of the total plasma area to the total image area (Equation (7)). SI takes in consideration the upper and lower limits (maximum and lower PR values) and calculates the percentage of change in that time interval. Evidently, the higher the value of SI the more susceptible is a color channel to change. SI follows the formula:

$$SI = \frac{|\min(V_{ch}) - \max(V_{ch})\,|}{\max(V_{ch})} \times 100 \quad (6)$$

V_{ch} is the value of a color channel at a specific objective magnification, and at a specific point of interest (maximum or the minimum value).

Figure 3 shows a flowchart representing an overview of the algorithm. Initially the input data, an AVI file, is preprocessed in three steps: adjustment of the frame rate (reduction needed fps), define and crop the region of interest, and separate the resulted colored images into their red, green, blue and gray components. Then, the first statistical analysis is performed to produce relevant indices and other parameters (i.e., threshold values for binarization, and normalization). At the second part (second row in the flowchart) the previously processed images are converted into binary, and various statistics are calculated. Finally, in the last part of the algorithm (last row of the flowchart), Equations (2)–(5) are implemented to derive the four new regions in each binary channel and to calculate the relevant statistics and indices.

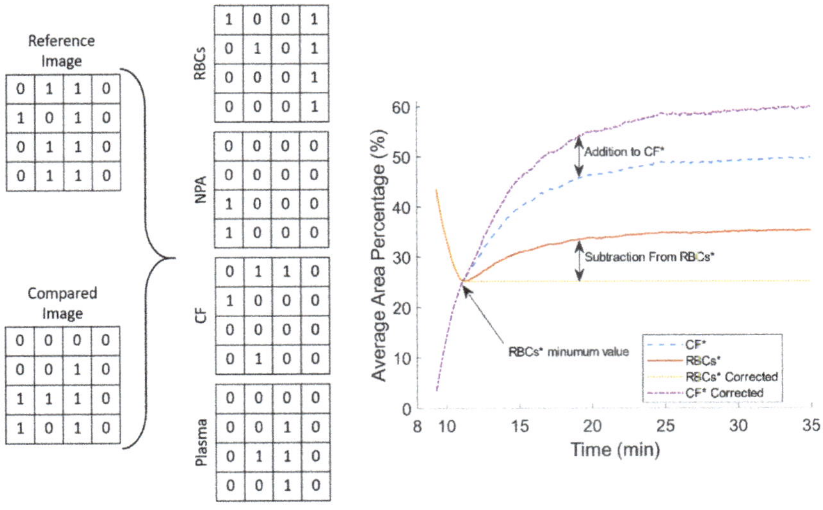

(a) (b)

Figure 2. (**a**) Representative binary comparison between parts of a reference and a subsequent image, and the resulted output matrices. (**b**) The behavior of *RBCs* * and *CF* * after clotting start (CS) and the correction process.

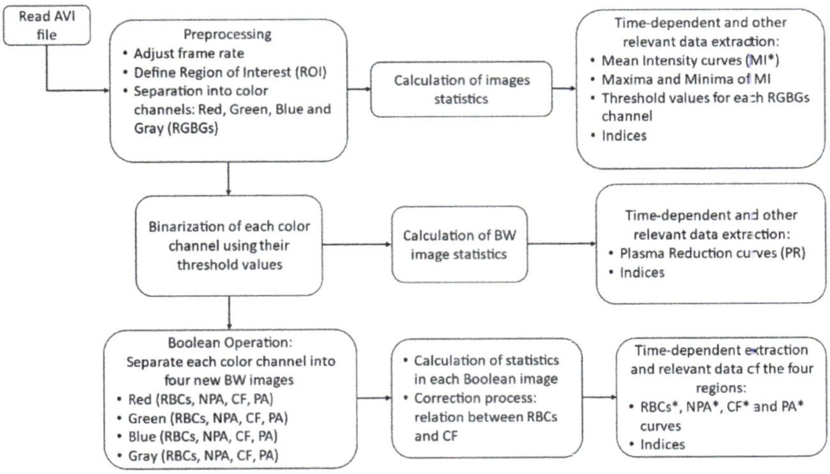

Figure 3. Flowchart of the proposed algorithms showing the three stages.

Figure 4 below shows the result of the processing algorithm for the 20× objective and for the gray level images, as the representative of the cases. As it can be seen the clot formation is detected as negligible in the initial time (t = CS), following a substantial increase at the half time of the coagulation process (t = CT), and reaching a maximum at the end of the time (t = 35 min). The area occupied by the RBCs is decreasing in time as the clotted regions extent on the sample. As noted earlier the plasma area (not shown) is complementary to the RBC area at the early stages, as coagulation progresses however, the plasma and RBC areas are occupied by clots, resulting in the CF images.

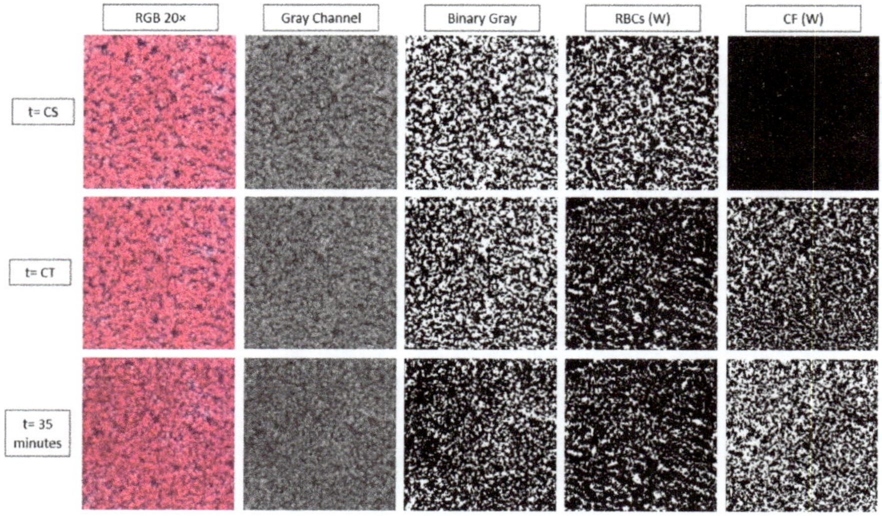

Figure 4. Representative images and the resulted red blood cell (RBC) and clot areas (CF) in white (W) pixels, on which the relevant indices are defined.

3. Results and Discussion

Three approaches have been utilized in this work for the whole blood coagulation assessment. In the first approach the mean intensity (MI) for each color channel was calculated for the three different magnification objectives. The images were separated into their four color channels in order to detect differences between them, regarding blood coagulation. The mean image intensity was examined in order to assess the overall effect of light blocking and the characteristics of whole blood coagulation time. The MI index for all color channels in Figure 5 has been normalized with its maximum value (and denoted as MI^*) in order to compare the behaviour of the different color channels. In this figure the time response of the RGBGs channels at three different magnifications is illustrated for the whole experimental period. In all cases, there is an increase in the average detected light until approximately the 10th minute, most probably due to the phenomenon of RBCA, in combination with the initial stages of fibrin formation [31]. After RBCA reaches a maximum and any cell movement has ceased, blood coagulation is the dominant process affecting blood microstructure. Indeed, after approximately 10 min from the peak of MI^*, there is a significant decrease in the MI curves as light is blocked due to the structural changes in the sample, caused by the formed clots.

MI^* curves appear in the same form for the specific magnifications, with some differences between the color channels. Differences are observed for the initial parts of the curves, however it should be noted that for each magnification a different test was performed. For example, in Figure 5a a spike like behavior around the second minute is observed, most likely due to a movement of blood components. In Figure 5c for the 50× objective, an obvious difference exists at the beginning of the process between the red and green color channels. In this magnification, small displacement of blood in the microchannel may have a greater effect in the results. After the maximum of the curves (which define the CS time), all curves have a similar behavior, with small differences between the experiments (i.e., the three different magnifications) and the color channels. The CS times observed in the present work for the non-reacted normal blood sample are in the vicinity ~10 min, agreeing with times observed in other studies. For reacted blood samples the onset of the coagulation process is detected at much smaller times depending on the reaction agent [16,35,36,38].

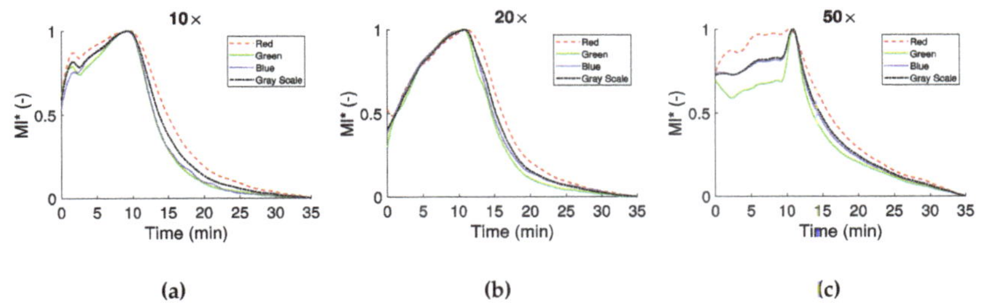

Figure 5. Normalized mean intensity MI^* of the RGBGs responses for the three different magnifications (10× in panel (**a**), 20× in panel (**b**) and 50× in panel (**c**)). The maximum value of MI is used for the normalization.

In the second approach the images were converted to binary, as described earlier. The binarization of the data resulted in the separation of the image into regions of plasma, RBCs and clots, and subsequently to the definition of the $RBCs^*$, NPA^*, CF^* and the PA^* indices. These indices were calculated for all consecutive frames from the CS time onward. As blood coagulates, and clots increase in number and size, the total number of pixels corresponding to plasma decrease could be assessed using the PR ratio:

$$PR = \frac{Total\ Plasma\ Area}{Total\ number\ of\ pixels} \tag{7}$$

PR is presented in Figure 6 for the 10×, 20× and 50× objectives, showing similar behavior between channels. This is expected, according to the MI^* curves in Figure 5a,b, which have a similar behavior after CS. Figure 6c shows a greater reduction in the initial value of PR, probably due to a lower threshold value in the processing stage caused by the higher objective magnification.

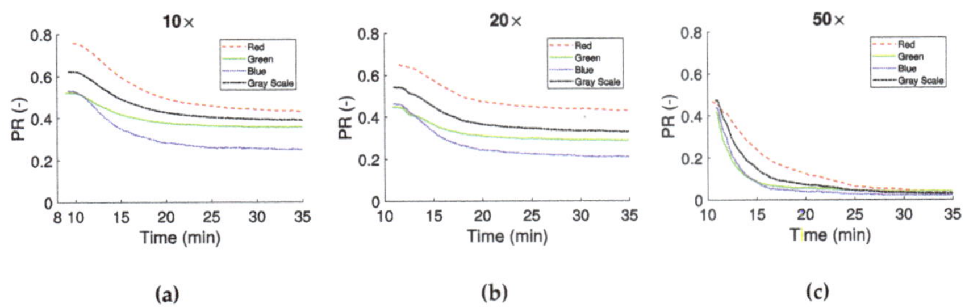

Figure 6. Plasma reduction (PR) as the ratio between the RBCs/clots and the plasma, of sequential images in time after CS. Panels (**a–c**) show the results from the 10×, 20× and 50× magnifications respectively.

Information on clot formation, provided by the CF^* index, is illustrated in Figure 7. Figure 7a,b correspond to 10× and 20× objectives respectively and show that CF^* (clot area in the image) has similar time response and maximum values at the steady state. Differences are apparent regarding the maximum values between each color channel at steady state. In the 50× objective case (Figure 7c) the shape of the curves for the color channels remained similar to the other two objectives, however the maximum CF^* values detected are higher. All CF^* curves have an exponential characteristic, and show a rapid growth of the forming clots in the early minutes after CS.

One of the objectives of this study was to identify the most efficient approach in terms of blood coagulation assessment using brightfield microscopy. Therefore, the results for MI^*, PR, and CF^* can be further analysed for sensitivity utilising the sensitivity index SI (Equation (6)). Tables 2–4 present the CS time, a characteristic coagulation time (CT), and SI for the three objectives used and for all color channels. CT is defined as the "half-life" time of the coagulation process after CS, i.e., the time required to reach the 50% of the CF^*

growth, or the 50% in *MI* * and *PR*, with maximum/minimum values defined at the 95% of the developed curves. The mean value and the standard deviation (SD) from the RGBGs channels are included for each index in the tables.

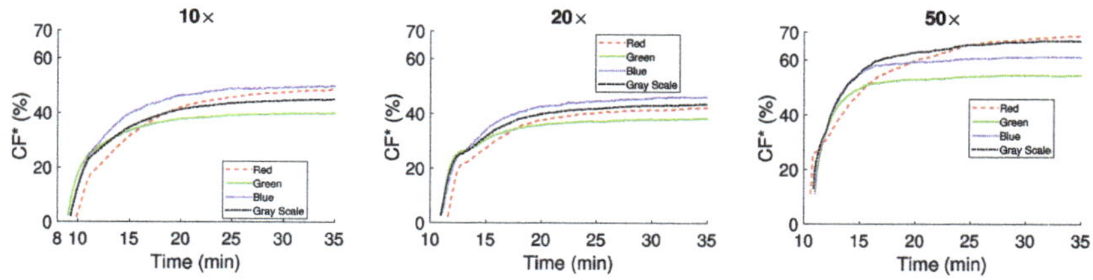

(a) (b) (c)

Figure 7. *CF* * expressed as percentage of the image area that is occupied by clots for the three magnifications (10× (**a**), 20× (**b**) and 50× (**c**)).

Table 2. Clot start time (CS), characteristic coagulation time (CT) and sensitivity index (SI) for mean intensity (*MI* *), plasma reduction (*PR*) and clot formation (*CF* *) in the 10 times objective case.

10×	*MI* *		*PR*		*CF*	
	CS(m:s)	CT(m:s)	CT(m:s)	SI(m:s)	CT(m:s)	SI(m:s)
Red	9:33	5:20	5:03	−40.54%	3:28	46.78%
Green	9:16	3:45	4:28	−28.20%	1:24	37.54%
Blue	9:05	4:00	4:20	−50.59%	2:12	46.15%
Gray	9:16	4:29	4:52	−34.41%	1:56	42.87%
Mean	9:18	4:24	4:41	−38.44%	2:15	43.34%
SD	±0:11	±0:42	±0:28	±9.54%	±0:52	±4.22%

Table 3. CS, CT and SI for mean intensity *MI**, plasma reduction *PR* and clot formation *CF* * in the 20.

20×	*MI* *		*PR*		*CF*	
	CS(m:s)	CT(m:s)	CT(m:s)	SI(m:s)	CT(m:s)	SI(m:s)
Red	11:05	5:00	4:23	−30.71%	1:43	39.79%
Green	10:41	3:53	3:43	−32.43%	0:51	35.17%
Blue	10:57	3:53	3:47	−52.33%	1:35	42.67%
Gray	10:53	4:20	4:19	−36.26%	1:19	40.93%
Mean	10:54	4:16	4:03	−37.93%	1:22	39.64%
SD	±0:10	±0:32	±0:20	±9.87%	±0:22	±3.20%

Table 4. CS, CT and SI for mean intensity *MI**, plasma reduction *PR* and clot formation *CF* * in the 50.

50×	*MI* *		*PR*		*CF*	
	CS(m:s)	CT(m:s)	CT(m:s)	SI (m:s)	CT(m:s)	SI(m:s)
Red	10:25	5:25	4:15	−92.45%	2:48	57.60%
Green	10:57	2:40	1:19	−90.36%	1:03	42.53%
Blue	10:53	3:24	1:35	−95.03%	1:11	50.20%
Gray	10:53	3:45	2:12	−93.57%	1:39	53.94%
Mean	10:46	3:48	2:20	−92.85%	1:40	51.07%
SD	±0:14	±1:09	±1:19	±1.96%	±0:47	±6.44%

Inspecting Tables 2–4 is apparent that the Mean CS times in the *MI* * analysis show good agreement, with values of 9:18 ± 0:11, 10:54 ± 0:10 and 10:46 ± 0:14 min and seconds, for the 10×, 20× and 50× objectives, respectively. Similarly, for all objectives the CT of the

MI * index shows similar Mean and SD values. Small differences in the Mean CS and CT values between the objectives could be attributed to the manual handling of the sample placement, and the sampling initiation time of the experiments. SD of the metrics CS, CT of *MI* *, and CT of *PR* are very close to each other for all the color channels in the individual experiments.

Despite the similarities in the general behavior of the indices, minor discrepancies are also apparent. For example, for the 10× and 20× objectives the Mean and SD of the SI for the *PR* index are −38.44 ± 9.54% and 37.93 ± 9.87%, respectively. For the 50× objective, however, this metric is equal to −92.85 ± 1.96%, a much higher mean value with smaller SD than the other objectives. This might be due to the higher magnification in the 50× lens, where there is a smaller sampling area, and therefore small cell movements, due to RBCA and/or the formation of new clots, have a greater effect in allowing/blocking the transmitted light. In the *CF* * analysis, the difference in the SI values between the 10× and 20× with the 50× objective is smaller than previously observed, with Mean and SD values of 43.34 ± 4.22%, 39.64 ± 3.20% and 51.07 ± 6.44% for the 10×, 20× and 50× objectives, respectively.

The results of the present study are comparable with those from recent and earlier works exploiting the optical properties of blood coagulation. These works include the laser speckle contrast imaging (LSCI) methods [41], which are also non-contact techniques. LSCI methods quantify light scattering and propagation, using wavelengths in the infrared/near infrared light spectrum [36–38]. These studies have established an exponential behavior of the assessed coagulation indices, with characteristics depending on reagent and sample used. This exponential behavior is observed in the present study for the non-activated coagulation of blood in Figure 7a–c. Moreover, the present results agree both qualitatively and quantitatively with data obtained using TEG instruments, in which CS times of approximately 12 min and CT of approximately 4 min are observed [42].

A popular method of choice, in several of the aforementioned and other studies in the literature, for the extraction of plasma, or whole blood coagulation parameters, is the calculation of the (cross- or/and auto-) correlation coefficients in consecutive images. Correlation methods can quantify the similarities between two signals (images), however with a computational cost related to the number of arithmetic operations, usually scaling with the square of the sample number (image size and number of images). The total number of operations in the binary comparison used in the present study, scales directly with the sample number. The binary method utilized in this work results into specific regions of interest, depending on the desired complexity (four regions in the present case). The separation of the four regions requires a small amount of processing power due to low algorithmic complexity, as shown in the Equations (2)–(5). Also, the visible light source offers true color images that can be used to further analyze whole blood coagulation process using for example feature extraction algorithms.

4. Limitations and Sources of Uncertainty

The blood sample used in the study was provided by the same donor, therefore negligible differences in the sample are expected between the various tests. The repeatability of the technique is still to be addressed in detail, however, an indication for the robustness of the technique is given by the similar behavior of the indices, which have been obtained from different tests in each magnification. As mentioned earlier, some uncertainty is expected in the CS and CT times, due to sample handling and manipulation as the sample placement was performed manually. Environmental factors (e.g., room temperature and illumination, etc.) are not expected to have a significant effect, as they were controlled and consistent throughout the experiments. The in-house construction of the microchannels may have resulted in small geometrical differences, however no significant influence is expected in that aspect. Accurate focusing of the microscope on the sample may be an issue for the high magnification cases (50× objective), however by inspection no significant issues were identified.

In terms of image processing, the opaqueness of the sample and the probable highe concentration RBC stacks from the RBCA process, can decrease the overall light absorptio by the image sensor. In the absence of the standard test, these results can be compared wit data from literature [34,38]. Nevertheless, the efficiency of the image processing metho adopted in this work is reflected in the analysis of the results and more specifically in th SI data.

5. Conclusions

In the present study, indices for the assessment of coagulation in a drop of blood wer developed, and experiments were conducted in custom made microchannels, in order t demonstrate the validity and the potential of the approach. The different magnificatio lenses used showed major similarities and minor differences, regarding the behavior of th various indices. The coagulation process in whole blood was found to have an exponentia behavior in agreement with other studies in the literature. The developed methodolog presented in this work is very promising for application in small scale diagnostic de vices for the point-of-care. Specifically, the current setup (brightfield illumination/camer imaging/low magnification microscopy/simple channel) is ideal for the construction o inexpensive POC devices. Further, the direct statistical analysis of the images, in combi nation with the Boolean approach can result in reduced computational needs, therefor allowing for onboard processing in a POC device.

Author Contributions: Conceptualization, E.K.; methodology, E.K.; software, M.L. and E.K.; valida tion, M.L.; formal analysis, M.L. and E.K.; investigation, M.L.; resources, E.K.; data curation, M.L writing—original draft preparation, M.L.; writing—review and editing, E.K.; visualization, M.L supervision, E.K.; project administration, E.K.; funding acquisition, E.K. Both authors have read an agreed to the published version of the manuscript.

Funding: The Project EXCELLENCE/0918/0215 is co-financed by the European Regional Develop ment Fund and the Republic of Cyprus through the Research and Innovation Foundation.

Acknowledgments: This study was supported from the Cyprus Research and Innovation Foundatio grant with ref.: EXCELLENCE/0918/0215. The help of Dimitris Pasias and Andreas Passos is als gratefully acknowledged.

Conflicts of Interest: The authors declare no conflict of interest.

References

1. Dahlbäck, B. Blood coagulation. *Lancet* **2000**, *355*, 1627–1632. [CrossRef]
2. Ruggeri, Z.M. The role of von Willebrand factor in thrombus formation. *Thromb. Res.* **2007**, *120*. [CrossRef]
3. Gale, A.J. Continuing Education Course #2: Current Understanding of Hemostasis. *Toxicol. Pathol.* **2011**, *39*, 273–280.
4. Weisel, J.W. The mechanical properties of fibrin for basic scientists and clinicians. *Biophys. Chem.* **2004**, *112*, 267–276. [CrossRef]
5. Lippi, G.; Favaloro, E.J.; Franchini, M.; Guidi, G.C. Milestones and perspectives in coagulation and hemostasis. *Semin. Thromb Hemost.* **2009**, *35*, 9–22. [CrossRef] [PubMed]
6. Alzahrani, S.H.; Ajjan, R.A. Review article: Coagulation and fibrinolysis in diabetes. *Diabetes Vasc. Dis. Res.* **2010**, *7*, 260–273. [CrossRef
7. Franchini, M.; Mannucci, P.M. Venous and arterial thrombosis: Different sides of the same coin? *Eur. J. Intern. Med.* **2008**, *19* 476–481. [CrossRef]
8. Federici, A.B. Prophylaxis of bleeding episodes in patients with von Willebrand's disease. *Blood Transfus.* **2008**, *6* (Suppl. 2), s26
9. Baskurt, O.K.; Meiselman, H.J. Erythrocyte aggregation: Basic aspects and clinical importance. *Clin. Hemorheol. Microcirc.* **2013** *53*, 23–37. [CrossRef]
10. Smith, S.A.; Travers, R.J.; Morrissey, J.H. How it all starts: Initiation of the clotting cascade. *Crit. Rev. Biochem. Mol. Biol.* **2015**, *50* 326–336. [CrossRef] [PubMed]
11. Mohammadi Aria, M.; Erten, A.; Yalcin, O. Technology Advancements in Blood Coagulation Measurements for Point-of-Car Diagnostic Testing. *Front. Bioeng. Biotechnol.* **2019**, *7*, 395. [CrossRef]
12. Paniccia, R.; Priora, R.; Liotta, A.A.; Abbate, R. Platelet Function tests: A Comparative Review. *Vasc. Health Risk Manag.* **2015**, *11* 133–148. [CrossRef]
13. Monroe, D.M.; Hoffman, M. What does it take to make the perfect clot? *Arterioscler. Thromb. Vasc. Biol.* **2006**, *26*, 41–48. [CrossRef]
14. Harris, L.F.; Castro-López, V.; Killard, A.J. Coagulation monitoring devices: Past, present, and future at the point of care *TrAC-Trends Anal. Chem.* **2013**, *50*, 85–95. [CrossRef]

15. Poller, L.; Ibrahim, S.; Keown, M.; Pattison, A.; Jespersen, J. The prothrombin time/international normalized ratio (PT/INR) Line: Derivation of local INR with commercial thromboplastins and coagulometers—Two independent studies. *J. Thromb. Haemost.* **2011**, *9*, 140–148. [CrossRef] [PubMed]
16. Lei, K.F.; Chen, K.H.; Tsui, P.H.; Tsang, N.M. Real-Time Electrical Impedimetric Monitoring of Blood Coagulation Process under Temperature and Hematocrit Variations Conducted in a Microfluidic Chip. *PLoS ONE* **2013**, *8*. [CrossRef]
17. Andrew, M.; Ansell, J.L.; Becker, D.M.; Becker, R.C.; Triplett, D.A. Point-of-care prothrombin time measurement for professional and patient self-testing use: A multicenter clinical experience. *Am. J. Clin. Pathol.* **2001**, *115*, 288–296. [CrossRef]
18. Nam, M.H.; Kyoung, H.R.; Pak, H.N.; Chang, K.L.; Kim, Y.H.; Kap, N.L.; Cho, Y. Evaluation of the Roche CoaguChek XS handheld coagulation analyzer in a cardiac outpatient clinic. *Ann. Clin. Lab. Sci.* **2008**, *38*, 37–40.
19. Hartert, H. Blutgerinnungsstudien mit der Thrombelastographie, einem neuen Untersuchungsverfahren. *Klin. Wochenschr.* **1948**, *26*, 577–583. [CrossRef] [PubMed]
20. Whiting, D.; Dinardo, J.A. TEG and ROTEM: Technology and clinical applications. *Am. J. Hematol.* **2014**, *89*, 228–232. [CrossRef]
21. Nardi, G.; Agostini, V.; Rondinelli, B.; Russo, E.; Bastianini, B.; Bini, G.; Bulgarelli, S.; Cingolani, E.; Donato, A.; Gambale, G.; et al. Trauma-induced coagulopathy: Impact of the early coagulation support protocol on blood product consumption, mortality and costs. *Crit. Care* **2015**, *19*. [CrossRef] [PubMed]
22. Schöchl, H.; Nienaber, U.; Hofer, G.; Voelckel, W.; Jambor, C.; Scharbert, G.; Kozek-Langenecker, S.; Solomon, C. Goal-directed coagulation management of major trauma patients using thromboelastometry (ROTEM®)-guided administration of fibrinogen concentrate and prothrombin complex concentrate. *Crit. Care* **2010**, *14*, 1–11. [CrossRef] [PubMed]
23. Davies, E.R. *Computer and Machine Vision*; Academic Press: Cambridge, MA, USA, 2012; ISBN 9780123869081.
24. Sotiropoulou, C.L.; Voudouris, L.; Gentsos, C.; Demiris, A.M.; Vassiliadis, N.; Nikolaidis, S. Real-time machine vision FPGA implementation for microfluidic monitoring on lab-on-chips. *IEEE Trans. Biomed. Circuits Syst.* **2014**. [CrossRef]
25. Chang, H.; Yang, Q.; Parvin, B. Segmentation of heterogeneous blob objects through voting and level set formulation. *Pattern Recognit. Lett.* **2007**. [CrossRef] [PubMed]
26. Kaliviotis, E.; Sherwood, J.M.; Balabani, S. Partitioning of red blood cell aggregates in bifurcating microscale flows. *Sci. Rep.* **2017**, *7*. [CrossRef]
27. Martínez-López, J.I.; Betancourt, H.A.; García-López, E.; Rodriguez, C.A.; Siller, H.R. Rapid fabrication of disposable micromixing arrays using xurography and laser ablation. *Micromachines* **2017**, *8*, 144. [CrossRef]
28. Roggo, C.; van der Meer, J.R. Miniaturized and integrated whole cell living bacterial sensors in field applicable autonomous devices. *Curr. Opin. Biotechnol.* **2017**, *45*, 24–33. [CrossRef]
29. Rosado-Galindo, H.; Domenech, M. Polystyrene Topography Sticker Array for Cell-Based Assays. *Recent Prog. Mater.* **2020**, *2*, 1–18. [CrossRef]
30. Pasias, D.; Passos, A.; Constantinides, G.; Balabani, S.; Kaliviotis, E. Surface tension driven flow of blood in a rectangular microfluidic channel: Effect of erythrocyte aggregation. *Phys. Fluids* **2020**. [CrossRef]
31. Zhukov, E.L.; Gubarev, F.A.; Tyutrin, I.I. A Portable Device for Rapid Assessment of the Functional State of the Hemostasis System. *Biomed. Eng.* **2021**, *54*, 345–349. [CrossRef]
32. Weissleder, R. A clearer vision for in vivo imaging: Progress continues in the development of smaller, more penetrable probes for biological imaging. *Nat. Biotechnol.* **2001**, *19*, 316–317. [CrossRef] [PubMed]
33. Panteleev, M.A.; Ovanesov, M.V.; Kireev, D.A.; Shibeko, A.M.; Sinauridze, E.I.; Ananyeva, N.M.; Butylin, A.A.; Saenko, E.L.; Ataullakhanov, F.I. Spatial propagation and localization of blood coagulation are regulated by intrinsic and protein C pathways, respectively. *Biophys. J.* **2006**, *90*, 1489–1500. [CrossRef] [PubMed]
34. Lim, H.; Nam, J.; Xue, S.; Shin, S. Measurement of blood coagulation with considering RBC aggregation through a microchip-based light transmission aggregometer. *Clin. Hemorheol. Microcirc.* **2011**, *47*, 211–218. [CrossRef] [PubMed]
35. Xu, X.; Zhu, J.; Chen, Z. Dynamic and quantitative assessment of blood coagulation using optical coherence elastography. *Sci. Rep.* **2016**, *6*. [CrossRef] [PubMed]
36. Faivre, M.; Peltié, P.; Planat-Chrétien, A.; Cosnier, M.-L.; Cubizolles, M.; Nougier, C.; Negrier, C.; Pouteau, P.; Négrier, C. Society of Photo-optical Instrumentation Engineers. *J. Biomed. Opt.* **2011**, *16*, 57001. [CrossRef]
37. Patiño-Velasco, M.M.; Andrade-Eraso, C.; Vásquez-López, J.; Trivi, M.; Rabal, H.J. Blood coagulation measurements using dynamic speckle technique. In Proceedings of the VI Latin American Congress on Biomedical Engineering CLAIB 2014, Paraná, Argentina, 29–31 October 2014; Volume 49, pp. 91–94.
38. Tripathi, M.M.; Hajjarian, Z.; Van Cott, E.M.; Nadkarni, S.K. Assessing blood coagulation status with laser speckle rheology. *Biomed. Opt. Express* **2014**, *5*, 817. [CrossRef]
39. Araújo, A.C.O.; Domingues, R.B.; van Bellen, B. Determinação do INR: Comparação entre método convencional e dispositivo portátil. *J. Vasc. Bras.* **2014**, *13*, 88–93. [CrossRef]
40. Savva, G.; Manousakis, K.; Ellinas, G. Eavesdropping-Aware Routing and Spectrum/Code Allocation in OFDM-Based EONs Using Spread Spectrum Techniques. *J. Opt. Commun. Netw.* **2019**, *11*, 409–421. [CrossRef]
41. Boas, D.A.; Dunn, A.K. Laser speckle contrast imaging in biomedical optics. *J. Biomed. Opt.* **2010**, *15*, 011109. [CrossRef]
42. Durila, M.; Kalincik, T.; Cvachovec, K.; Filho, R. Heparinase-modified thromboelastography can result in a fibrinolytic pattern. *Anaesthesia* **2010**, *65*, 864–865. [CrossRef]

biosensors

MDPI

Article

Simultaneous Imaging of Bio- and Non-Conductive Targets by Combining Frequency and Time Difference Imaging Methods in Electrical Impedance Tomography

Xue Bai [1], Dun Liu [2,*], Jinzhao Wei [1], Xu Bai [1], Shijie Sun [1,2] and Wenbin Tian [3]

1 School of Instrumentation and Optoelectronic Engineering, Beihang University, Beijing 100191, China; bxbuaabj@buaa.edu.cn (X.B.); weijinzhao@buaa.edu.cn (J.W.); xubai11@163.com (X.B.); sunsj@buaa.edu.cn (S.S.)
2 Beijing Advanced Innovation Center for Big Data-Based Precision Medicine, Beihang University, Beijing 100191, China
3 College of Engineering, China Agricultural University, Beijing 100083, China; wenbin.tian@hotmail.com
* Correspondence: liudun@buaa.edu.cn

Abstract: As a promising medical imaging modality, electrical impedance tomography (EIT) can image the electrical properties within a region of interest using electrical measurements applied at electrodes on the region boundary. This paper proposes to combine frequency and time difference imaging methods in EIT to simultaneously image bio- and non-conductive targets, where the image fusion is accomplished by applying a wavelet-based technique. To enable image fusion, both time and frequency difference imaging methods are investigated regarding the reconstruction of bio- or non-conductive inclusions in the target region at varied excitation frequencies, indicating that none of those two methods can tackle with the scenarios where both bio- and non-conductive inclusions exist. This dilemma can be resolved by fusing the time difference (td) and appropriate frequency difference (fd) EIT images since they are complementary to each other. Through simulation and in vitro experiment, it is demonstrated that the proposed fusion method can reasonably reconstruct both the bio- and non-conductive inclusions within the lung models established to simulate the ventilation process, which is expected to be beneficial for the diagnosis of lung-tissue related diseases by EIT.

Keywords: electrical impedance tomography; frequency difference; time difference; lung imaging

Citation: Bai, X.; Liu, D.; Wei, J.; Bai, X.; Sun, S.; Tian, W. Simultaneous Imaging of Bio- and Non-Conductive Targets by Combining Frequency and Time Difference Imaging Methods in Electrical Impedance Tomography. *Biosensors* **2021**, *11*, 176. https://doi.org/10.3390/bios11060176

Received: 20 April 2021
Accepted: 28 May 2021
Published: 31 May 2021

1. Introduction

Electrical impedance tomography (EIT) is a non-invasive imaging technique in which the absolute value or the change of conductivity distributions are reconstructed [1–6]. The underlying rotational is based on generating a current density distribution inside the medium and measuring the resultant electromagnetic fields by means of sensors positioned circumferentially on the boundary [2]. Current excitation and voltage acquisition are repeated continuously and rapidly until all the independent electrode combinations are deployed. Attributed to its notable merits of high temporal resolution, non-radiation, low cost, and non-intrusive measurement, EIT has also gained a great deal of interest in monitoring rapid changes in multiphase flow and medical imaging [2–6].

Ever since medical interest in EIT increased in 1980s, EIT has been developed to be a promising medical imaging technique to attain useful diagnostic information where there are obvious conductive changes in the biological tissues and organs with changes in physiological conditions, such as breathing, blood flow, digestive, canceration, and nervous activity [7,8]. Up to now, the medical applications of EIT research have mainly focused on thoracic imaging. Among the organs contained in the thorax, large volumes of air are Amoved into and out of the lungs, leading to obvious conductivity changes during ventilation. Therefore, it is agreed that monitoring of breathing by imaging the lung is the most important physiological application of this technique. Monitoring ventilation by EIT has already been applied for managing mechanically ventilated patients as it is can guide the set-up of mechanical ventilator pressure, volume, and respiratory rate [9].

Most clinical and physiological research into the lung with EIT is conducted using the difference imaging method to obtain air distribution images of ventilation and perfusion in the chest. Static imaging, i.e., reconstructing the absolute conductivity distribution of interest from a single set of measurement, has already been abandoned because of the poor image quality. Time difference EIT (td-EIT) was the consensus on the performance figures of merits for EIT image reconstruction after it was proposed by Barber and Brown [10]. Meanwhile, it was tested successfully for lung imaging as it can ameliorate modeling errors contained in the measurement data [11]. By taking the merits of frequency dependence of biological tissue, frequency difference EIT (fd-EIT) aroused significant interest by injecting currents with two or more different frequencies [11–18]. Compared with the td-EIT imaging method, fd-EIT does not need the reference measurement from the past, so it tackles the issue in terms of the unavailable voltage data with respect to the reference measurement in a real clinic environment [11–13]. Subsequently, developing a multi-frequency EIT (mf-EIT) system is an important transition for the research of EIT. Remarkable work includes the Sheffield system [14], the UCL systems [15], the KHU systems [16,17], and the Edinburgh system [3].

Many lung pathologies, such as atelectasis (lung collapse), pneumothorax, pleural effusion, and pulmonary edema, explicitly indicate an abnormity of air or liquid content in the lung. For example, the "pleural effusion" model in EIT contains a high conductivity region due to the accumulation of liquid in the pleural cavity [9,18]. Generally, diagnosis of several lung pathologies, such as avoiding acute respiratory distress syndrome (ARDS) with assessment of extravascular lung water, which is also known as quantify pulmonary edema and which can be considered as a biomaterial, may benefit from the use of the fd-EIT imaging method [19]. However, fd-EIT imaging faces a main challenge. Biomaterials such as lung tissues are only shown in the reconstructed images as their conductivity varies with frequency of the injected current. Thus, for lung imaging, the reconstructed images of fd-EIT are not well characterized in terms of attaining explicit interpretation of the pulmonary monitoring and air content. For the td-image reconstruction, it can be expected that the biological materials would induce much smaller conductivity variation than that of the non-conductive materials, and thus the biomaterial distribution would be hardly seen in the reconstructed images due to the existence of the large amount of air. Another limitation is that most EIT signals from ventilation may overwhelm those from perfusion. Therefore, EIT on lung monitoring by using only fd-EIT or td-EIT cannot stratify certain clinical requirements. In medical imaging applications, the images of different modalities have already been fused to obtain reliable and accurate diagnoses [20]. Inspired by the image fusion concept, it is meaningful to investigate the possibility of improving the image quality by fusing the reconstructed images of the td-EIT and fd-EIT since it is convenient to acquire both the td-EIT and fd-EIT data simultaneously.

In this paper, to simultaneously image bio- and non-conductive targets by EIT, it is proposed to fuse the reconstructed images by fd- and td-EIT to reveal both the bio- and non-conductive materials distributions, where the wavelet-based fusion method is employed. At the beginning, the td- and fd-EIT imaging methods to image bio- or non-conductive material distributions are investigated. This provides an in-depth look at the limitations of td- and fd-EIT imaging. The wavelet transform was performed to decompose and fuse the reconstructed td- and fd-EIT images, with the performances of typical wavelet basis functions compared to attain optimized image quality. Since the inherent properties of the lungs during breathing are well matched to the application of the proposed method, the lung ventilation imaging was studied. For the lung imaging, a simulation study was carried out by considering a homogenous phantom of the human thorax with lungs simulated as real-shaped abnormal inclusions containing air. The proposed method was evaluated by imaging the simulated ventilation process of the lung. In the end, experiments were conducted to validate the proposed method by adopting an in vitro lung model made with a white gourd.

This paper is structured as follows: In Section 2, the image reconstruction in EIT is briefly reviewed, and the wavelet-based fusion method for td- and fd-EIT images is demonstrated. In Sections 3 and 4, both simulation and experimental results are presented to validate the proposed method with simulated or in vitro lung models. Finally, conclusions are drawn in Section 5.

2. Methodology

2.1. Principles of EIT

In this investigation the boundary was $\partial\Omega$. Generally, for computational needs, it is assumed that Ω represents the body under the finite element model (FEM), Ω is discretized into small pixels or voxels and assigned a piecewise constant conductivity distribution, i.e., $\sigma = [\sigma_1, \ldots, \sigma_N]^T \in \mathbb{R}^N$ is assumed within Ω. The electric potential inside the sensing area induced by the injected AC current can be written as [5,6]:

$$\nabla \cdot (\sigma(x,y)\nabla u(x,y)) = 0,\ (x,y) \in \Omega \quad (1)$$

where u is the electric potential inside Ω. EIT is a nonlinear problem as the excitation current is a function of unknown conductivity distribution [21,22]. Therefore, based on the nonlinear model, the relation between the conductivity inside Ω and the boundary voltages at the boundary, $\partial\Omega$, can be expressed as:

$$V = U(\sigma) + e \quad (2)$$

where V is the measured voltage, U is the forward operator, and e is the additive noise and measurement error. Regarding the difference EIT imaging, Equation (2) can be linearized given a small conductivity perturbation $\Delta\sigma$, which can be expressed as (3):

$$\Delta V \approx J\Delta\sigma \quad (3)$$

where J is the Jacobian matrix, and calculated by solving the EIT forward problem, i.e., Equation (1). Difference EIT calculates a vector of conductivity changes between the measured conductivity distribution and the reference conductivity distribution. One commonly employed method is the single frequency td-EIT imaging. More specifically, the AC current with selected frequency is injected into the sensing area. Hereafter, a measurement at time t_0 is acquired as the reference, and then the voltage change at time t_1 against the reference measurement is employed for image reconstruction, which is expressed as:

$$\Delta V = V_{t_1} - V_{t_0} \quad (4)$$

In most cases, a set of reference measurements is acquired when the EIT sensor is filled with the background medium. To improve its precision, the measured data is typically averaged over many data frames.

Recently, incorporating the advances that have been made on the hardware system, a more attractive method in biomedical applications is the frequency-difference (fd-) EIT imaging. For the fd-EIT imaging, the current source is required to output a pair of complementary currents with two different frequency components, i.e., f_1 and f_2, which are injected to the sensor simultaneously or separately. At each frequency component, the boundary voltages are measured. The difference between the two sets of measured voltages is adopted for image reconstruction, which is calculated by

$$\Delta V = V_{f_2} - V_{f_1} \quad (5)$$

After attaining the voltage difference, to solve the inverse or image reconstruction problem in EIT, a variety of algorithms have been proposed in the literature [23]. In general, using the regularization technique is the predominant way. The conductivity distribution can be obtained by solving the optimization problem:

$$arg\ min\left\{\frac{1}{2}\parallel J\Delta\sigma - \Delta V \parallel_2^2 + \lambda R(\Delta\sigma)\right\} \tag{6}$$

where R is the regularization function incorporating a priori knowledge and λ is the regularization parameter. In this case, for lung EIT, traditional and proprietary algorithms have been an obstacle to interpretation of EIT images as the reconstructed images cannot reveal the character of the lung [2]. GREIT (Graz consensus reconstruction algorithm for EIT) is a type of regularized imaging approach for 2D linear EIT reconstruction of the lung, which is based on the experience with a large number of linear regularized EIT reconstructions. One of the advantages of GREIT is that it conforms to the performance requirements for lung application, while other reconstruction algorithms do not. One can refer to Adler [2] for the detailed description of GREIT.

2.2. Wavelet-Based Image Fusion

Image fusion has three levels: pixel level, feature level, and decision level [20]. Pixel-level fusion transcends other image fusion levels as it retains more detailed image information and can achieve higher accuracy. Among the pixel-level fusion methods, the wavelet-based method is recommended because of its non-redundancy and directionality [24–27]. Figure 1 shows the image fusion process based on the wavelet transform.

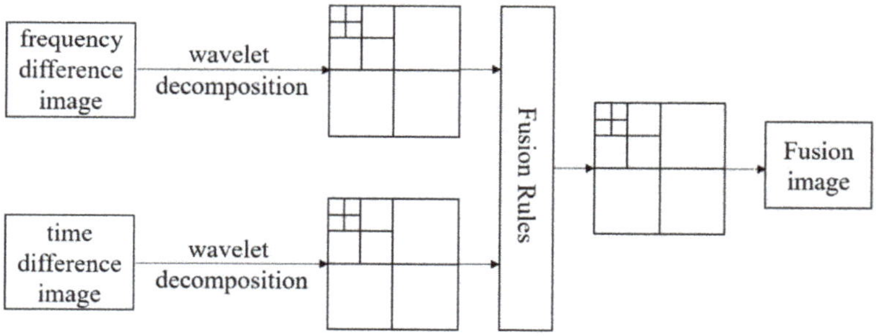

Figure 1. Image fusion process based on wavelet transform.

The fusion process is performed as follows: Firstly, the source images were decomposed by applying the wavelet transform with an appropriate wavelet basis function. In this procedure, reconstructed images by fd- and td-EIT can be assumed as a continuous change of two-dimensional signal $f(x_1, x_2) \in L^2(R^2)$, x_1 and x_2 are its abscissa and ordinate, respectively. The wavelet transform of $f(x_1, x_2)$ is expressed as:

$$WT_f(a, b_1, b_2) = \langle f(x_1, x_2), \psi_{a,b_1,b_2}(x_1, x_2)\rangle \tag{7}$$

$$\psi_{a,b_1,b_2}(x_1, x_2) = \frac{1}{a}\psi(\frac{x_1 - b_1}{a}, \frac{x_2 - b_2}{a}) \tag{8}$$

where $\psi(x_1, x_2)$ is a two-dimensional wavelet basis function, a is the scaling factor, and b is the displacement factor. By applying the basis function, the low-frequency and high-frequency components of the reconstructed image can be obtained, respectively. The low-frequency components of the image give approximate features, while the high-frequency components provide detailed features [27]. Secondly, appropriate fusion rules are employed to fuse the high frequency and low frequency components. The choices of fusion rules are critical in this procedure. There are many fusion rules, such as larger absolute value of coefficients, weighted average method, and local variance criterion [23]. According to previous studies, it is suggested that the fusion rules should adapt to the features of the source images. In this study, the fused image needs to maintain the position, size, edge, and shape of the target inclusions in the reconstructed images by fd- and td-EIT. In view of these requirements, the high frequency components were fused by taking the maximum absolute value in the source images. By combining the pixels correlation and regional variance in the image area, a hybrid fusion method was adopted for the fusion of low frequency components, as introduced by Wang et al. [27].

In the hybrid fusion rule, assuming that $L(X)$ represents the low frequency coefficient matrix of the image X, let $G(X,\ p)$ denote the regional variance of the region Q centred at p(m, n) in the low-frequency coefficient matrix of image X, which can be expressed as [27]:

$$G(X,p) = \sum_{q \subset Q} w(q)|L(X,p) - \overline{u}(X,p)|^2 \tag{9}$$

where $L(X,p)$ represents the value of the element at (m, n) in the low-frequency component coefficient matrix, and $\overline{u}(X,p)$ represents the average value of the elements inside Q. $w(q)$ is the weight coefficient, which would increase if it approached p(m, n). $M_2(p)$ is the matching degree of the regional variance regarding the low-frequency coefficient matrix of the source images A and B at p(m, n), which is defined as:

$$M_2(P) = \frac{2\sum_{q \subset Q} w(q)|L(A,p) - \overline{u}(A,p)||L(B,p) - \overline{u}(B,p)|}{G(A,p) + G(B,p)} \tag{10}$$

It is noted that the value of $M_2(p)$ varies from 0 to 1. The smaller the $M_2(p)$, the lower the correlation between the low frequency coefficient matrices. Let $t_2(0.5 < t_2 < 1)$ be the threshold of the matching degree. If $M_2(p) < t_2$, the option fusion strategy is adopted:

$$L(F,p) = \begin{cases} L(A,p), & G(A,p) \geq G(B,p) \\ L(B,p), & G(A,p) < G(B,p) \end{cases} \tag{11}$$

Otherwise, the average fusion strategy is used:

$$L(F,p) = \begin{cases} W_{max}L(A,p) + W_{min}L(B,p), & G(A,p) \geq G(B,p) \\ W_{min}L(A,p) + W_{max}L(B,p), & G(A,p) < G(B,p) \end{cases} \tag{12}$$

$$W_{min} = 0.5 - 0.5\left(\frac{1 - M_2(p)}{1 - t_2}\right) \tag{13}$$

$$W_{max} = 1 - W_{min} \tag{14}$$

2.3. Image Quality Evaluation

Quantitative measures, e.g., relative coverage ratio (*RCR*), relative image error (*RIE*), and structural similarity (*SSIM*) index are defined to evaluate the quality of reconstructed images, which are the differences and correlations between the true image and the reconstructed image [28]:

$$RCR = \frac{CR}{CR_{True}} \tag{15}$$

$$RIE = \frac{\| \hat{g} - g \|}{\| g \|} \tag{16}$$

$$SSIM = \frac{2\mu_x\mu_y \cdot 2\sigma_x\sigma_y}{\left(\mu_x^2 + \mu_y^2\right)\left(\mu_x^2 + \mu_y^2\right)} \tag{17}$$

where CR denotes the coverage ratio defined as the ratio of the size of the inclusions to the total size of the sensing area. Correspondingly, CR_{True} is CR of the true target distribution. g and $\hat{g}$ are the true and reconstructed images, respectively, μ_x and μ_y are the mean grey values in the true and reconstructed images, respectively, with σ_x and σ_y being the corresponding standard deviations, respectively. Quantitative evaluation of the reconstructed images of biological objects under different frequencies is challenging because the conductivity changes with frequencies and the ground truth is unknown especially for the fd-EIT imaging where different images can be obtained with different frequency intervals. Therefore, the popular criteria, such as image errors and correlation coefficients, cannot be employed in this case.

3. Characterization of td- and fd-EIT Imaging by Experiment

A series of static experiments were conducted to characterize the td- and fd-EIT imaging and validate the feasibility of the proposed method.

3.1. Experimental Setup

An experimental system with a cylindrical tank sensor was established to characterize the fd- and td-EIT imaging, as shown in Figure 2. The diameter of the tank was 38 cm with four layers of EIT electrodes mounted axially. In each layer, 32 metallic electrodes (stainless steel screws, radius = 5 mm) were uniformly mounted on the inner surface of the tank. Sixteen of them (choosing one in the other two electrodes) in the second layer were employed for EIT imaging. A National Instruments (NI) PXI-based EIT system developed by Beihang University was employed in the experiment. The system mainly comprises three modules: the AC current source, the data collection system, and the PC with system control functions for image reconstruction. In the experiment, the cylindrical tank was filled with tap water as the background medium with a conductivity of 0.034 S/m. The pairwise injected current had a peak-to-peak magnitude of around 1 mA, while the frequency was set to 1 KHz, 5 KHz, 10 kHz, 20 kHz, and 50 KHz. With the adjacent excitation strategy, the excitation current was injected into a pair of adjacent electrodes. By multiplexing, the potential differences between each possible pair of adjacent electrodes were measured by the data collection system. The measurements were then sent to the PC for image reconstruction via a USB interface. Finally, an image of the target inclusions was reconstructed in MATLAB using the data received. For a single measurement, 20 frames of data were averaged to mitigate the measurement noise.

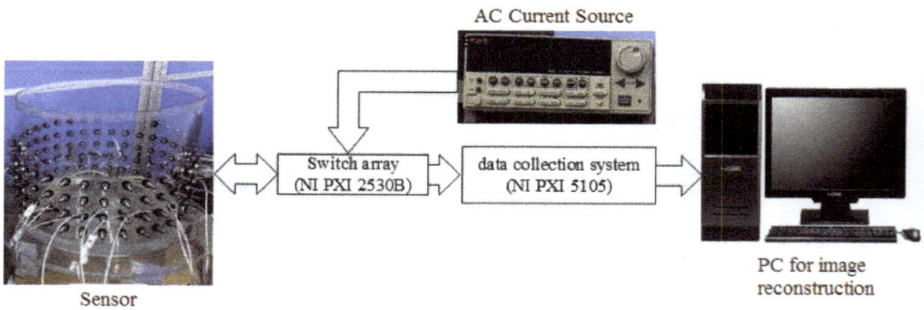

Figure 2. Experimental set-up.

Td- and fd-EIT were used to image three different phantoms, which were set up by inserting a conductive inclusion (a peeled radish in phantom 1), a non-conductive inclusion (a nylon rod in phantom 2), and both of them (in phantom 3) at a certain location in the tank. The diameters of the peeled radish and nylon rod were 60 mm and 55 mm, respectively. Their length was sufficiently long to extend beyond the water surface so that all tests were conducted with target inclusions homogeneous in the vertical direction. Figure 3a–f show the phantom pictures and corresponding distributions of the inclusions, respectively.

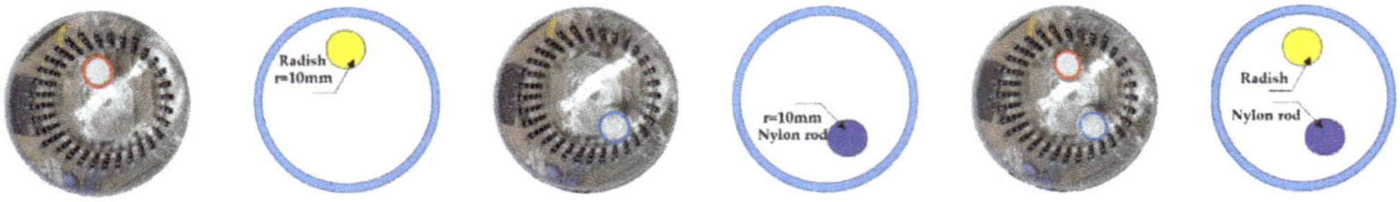

(a) (b) (c) (d) (e) (f)

Figure 3. Experimental setup of three different phantoms. (**a**) Phantom 1, peeled radish; (**b**) distribution of phantom 1; (**c**) phantom 2, nylon rod; (**d**) distribution of phantom 2; (**e**) phantom 3, peeled radish and nylon rod; and (**f**) distribution of phantom 3.

According to previous work [29,30], electrical impedance spectroscopy (EIS) can reveal the frequency response of electrical impedance of bio-tissues, which can be employed to demonstrate the frequency-dependencies of the conductivities of the peeled radish and nylon rod. Figure 4 shows the EIS measurements of the peeled radish and nylon rod by an impedance analyzer (MFLI Lock-in Amplifier-500 kHz/5 MHz, Zurich Instruments, Zurich, Switzerland) over the selected frequency range, i.e., 0–100 kHz. The results of EIS indicated that the conductivity of the radish increased linearly with the increase in frequency, and the linear fitting formula was:

$$\sigma = 0.000563 \times f + 0.02374 \tag{18}$$

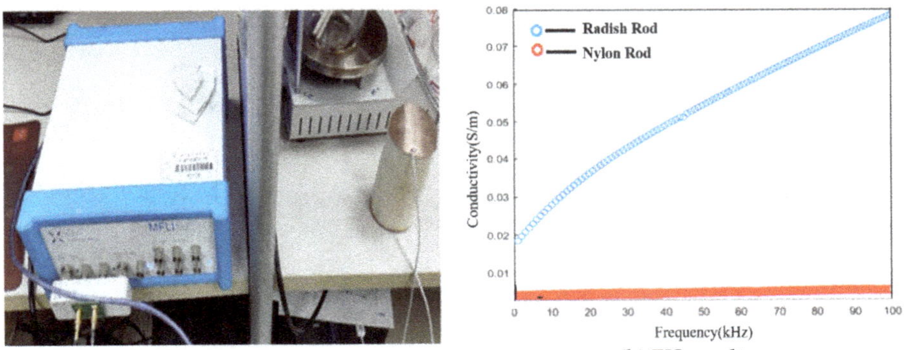

(**a**) EIS measurement (**b**) EIS results

Figure 4. Electrical impedance spectroscopy of peeled radish and nylon.

Its RMSE (root mean squared error) was 0.001834, and the value of R-square (coefficient of determination) was 0.9877.

In particular, it is noted that the conductivity of the radish at 10 kHz was about 0.03 S/m, which is nearly the same as the conductivity of the background medium. With respect to the nylon rod, the EIS data illustrate that its conductivity stayed almost unchanged with frequency. This is consistent with the previous analysis and verifies the feasibility of the use of the radish and nylon rod for the dedicated study on td- and fd-EIT imaging.

3.2. Experimental Results and Analysis

By applying the GREIT algorithms for image reconstruction, the td- and fd-EIT imaging results of the three phantoms were attained, and the size of the images w 64 × 64 pixels. In fd-EIT, 1 kHz was selected as the reference frequency. The voltage differences between the measured voltages at all others frequencies and at the reference frequency were procured for image reconstruction. In td-EIT, the measured voltage when the EIT sensor was filled with the background medium was taken as the reference, i.e., the measurement at t_1, and the voltage differences were obtained when perturbations were present in the background medium under the tested frequencies. The td- and fd-EIT imaging results are shown in Figures 5–7.

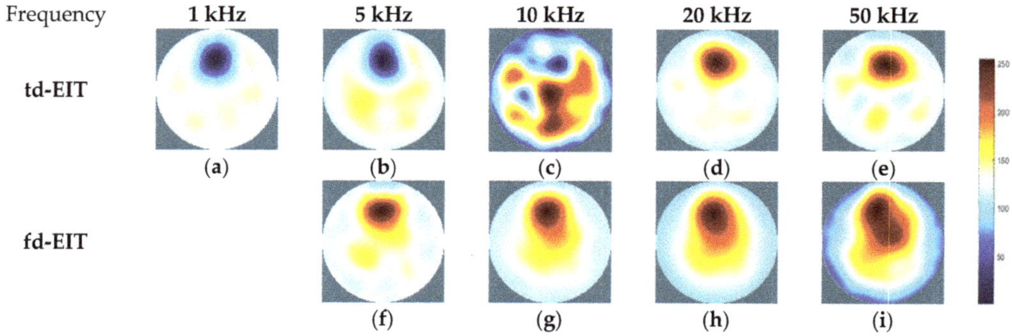

Figure 5. Td- and fd-EIT imaging results of phantom 1. (**a–e**) are td-EIT images of phantom 1 at 1, 5, 10, 20 and 50 kHz, respectively; (**f–i**) are fd-EIT images of phantom 1 at 5, 10, 20 and 50 kHz, respectively.

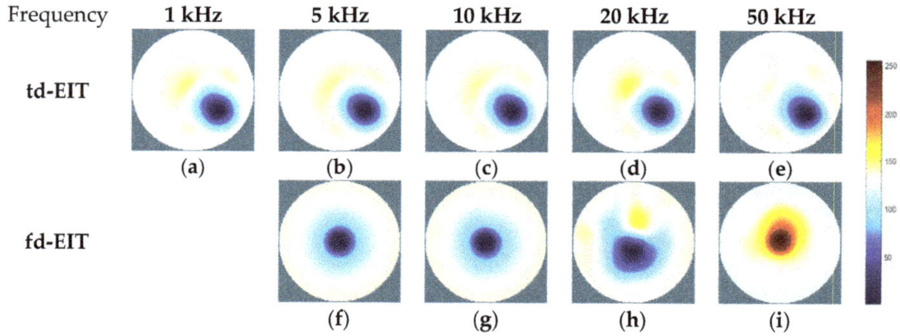

Figure 6. Td- and fd-EIT imaging results of phantom 2. (**a–e**) are td-EIT images of phantom 2 at 1, 5, 10, 20 and 50 kHz, respectively; (**f–i**) are fd-EIT images of phantom 2 at 5, 10, 20 and 50 kHz, respectively.

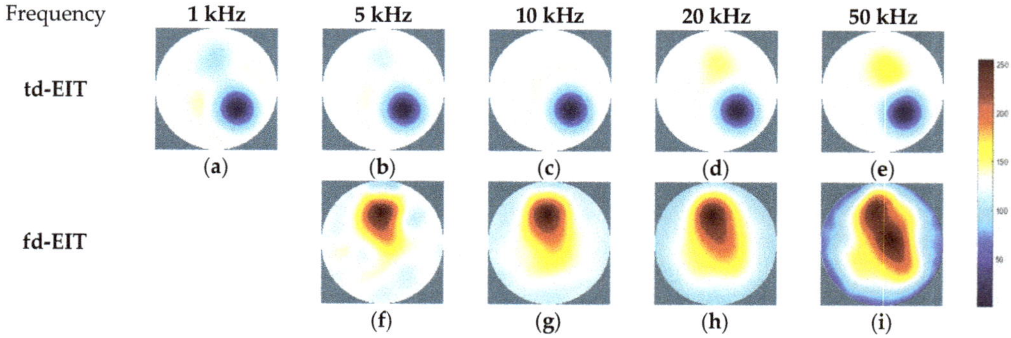

Figure 7. Td- and fd-EIT imaging results of phantom 3. (**a–e**) are td-EIT images of phantom 3 at 1, 5, 10, 20 and 50 kHz, respectively; (**f–i**) are fd-EIT images of phantom 3 at 5, 10, 20 and 50 kHz, respectively.

Figures 5 and 6 show the td- and fd- imaging results of phantoms 1 and 2 at 1 kHz, 5 kHz, 10 kHz, 20 kHz, and 50 kHz. Figures 5–7 have the same color bars, and the gray value range was 0–255, corresponding to the conductivity value of 0–0.08 S/m, respectively. The same applied to the relationship between the gray value and the conductivity in Figure 8, despite of the use of different colors for presentation. As indicated, td-EIT can reconstruct the position and shape of the single radish or nylon rob clearly (except for at 10 kHz where the conductivity of radish was close to the background medium as indicated by td-EIT imaging results). Fd-EIT could image the radish in phantom 1, but it failed to indicate the accurate position of the nylon rod in the reconstructed image since the conductivity of non-conductive nylon rod was not frequency-dependent. Regarding the phantoms 1 and 2, td-EIT provided better imaging results of the inclusions, as indicated in Tables 1 and 2 for quantitative evaluation, where td-EIT imaging results have lower RIE, as well as RCR and SSIM closer to 1. In Figure 5, td-EIT imaging results indicate that the color scale of the reconstructed inclusion changed with the increase in excitation frequency, which was due to the frequency-dependency of the conductivity of peeled radish. More specifically, the conductivity of peeled radish was lower than the tap water when the frequency was lower than 10 kHz, and larger than the tap water at frequencies >10 kHz. On the other hand, since the conductivity of the radish was related to frequency, the image quality was sensitive to the frequency intervals. Besides, size overestimation and image distortion are more significant with increase in the frequency intervals, which is attributed to the decrease of SNR with the increase in frequency. The image distortion was the most serious at 10 kHz due to the conductivity of peeled radish being very close to the tap water and the measured voltages had a much smaller signal-to-noise ratio (SNR), which would lead to wrong imaging result since the EIT reconstruction was ill-conditioned.

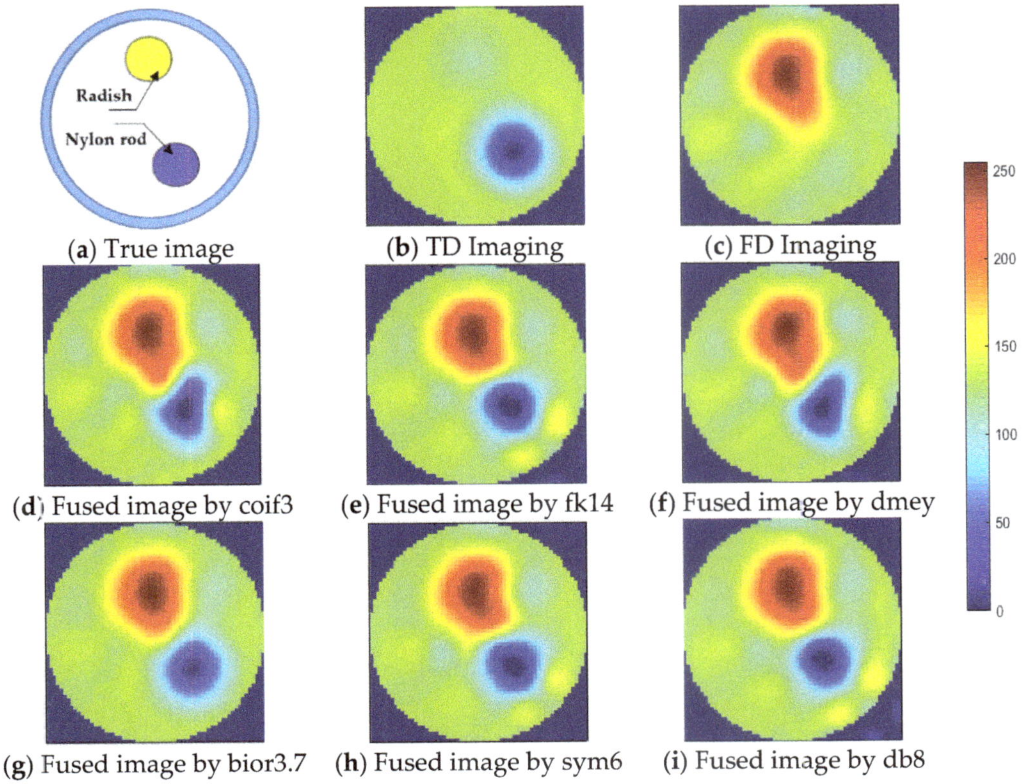

Figure 8. Fused images by applying the wavelet-based fusion method: (**a**) Ture image, (**b**) TD imaging, (**c**) FD imaging, (**d**) fused image by coif3, (**e**) fused image by fk14, (**f**) fused image by dmey, (**g**) fused image by bior3.7, (**h**) fused image by sym6, and (**i**) fused image by db8.

Table 1. RCR, RIE, and SSIM regarding td- and fd-EIT imaging results of phantom 1.

Frequency		1 kHz	5 kHz	10 kHz	20 kHz	50 kHz
RCR	fd-EIT	\	0.8309	1.1268	1.8621	4.0221
	td-EIT	0.7463	0.6415	\	1.1047	1.0386
RIE	fd-EIT	\	0.3933	0.4017	0.2950	0.4088
	td-EIT	0.0893	0.1156	\	0.2160	0.2496
SSIM	fd-EIT	\	0.7648	0.6764	0.4385	0.1614
	td-EIT	0.8039	0.7071	\	0.8583	0.6943

Table 2. RCR, RIE, and SSIM regarding td- and fd-EIT imaging results of phantom 2.

Frequency		1 kHz	5 kHz	10 kHz	20 kHz	50 kHz
RCR	fd-EIT	\	1.1667	1.1667	1.4118	1.0588
	td-EIT	1.0172	1.0221	1.0147	0.9706	1.0049
RIE	fd-EIT	\	0.1407	0.1421	0.1850	0.4487
	td-EIT	0.1017	0.0961	0.0964	0.1157	0.0882
SSIM	fd-EIT	\	-0.0272	−0.0272	−0.0284	−0.0262
	td-EIT	0.7544	0.7549	0.7578	0.7574	0.8169

For further demonstration, imaging results of phantom 3 by td- and fd-EIT are shown Figure 7. In Figure 7, the position and shape of the nylon rod can be well reconstructed by td-EIT, while due to the existence of the non-conductive nylon rod, the radish can hardly be seen as it causes much smaller responses at the sensor boundary. However, regarding fd-EIT imaging, only the peeled radish is shown in the reconstructed images. The reason is that the conductivity of the radish varies with excitation frequency while that of the nylon rod does not. These findings are verified by the evaluation metrics in Table 3.

Table 3. RCR, RIE, and SSIM regarding td- and fd-EIT imaging results of phantom 3.

Frequency		1 kHz	5 kHz	10 kHz	20 kHz	50 kHz
RCR	fd-EIT	\	1.0129	1.3462	2.1952	4.7882
	td-EIT	1.1007	1.1032	1.1106	1.1253	1.1327
RIE	fd-EIT	\	0.3862	0.4052	0.3079	0.4346
	td-EIT	0.0986	0.0869	0.0844	0.1096	0.1330
SSIM	fd-EIT	\	0.7832	0.7071	0.4318	0.1339
	td-EIT	0.8939	0.8998	0.8912	0.8930	0.8930

3.3. *Image Fusion with Respect to td- and fd-EIT Imaging*

To evaluate the performance of the above-mentioned fusion strategy, the reconstructed images of phantom 3 by td- and fd-EIT at 5 kHz were taken as the source images. Meanwhile, to investigate the impact of wavelet basis function, various basis functions available in MATLAB were employed in the wavelet transform, including "coif3", "fk14", "bior3.7", "dmey", "sym6", and "db8".

As shown in Figure 8, the source images by td- and fd-EIT cannot either demonstrate the conductive inclusion (i.e., the radish) or the non-conductive inclusion (i.e., the nylon rod). By applying the wavelet-based image fusion strategy, the fused image can successfully recover both the conductive and non-conductive inclusions, which resolves the issues of td- and fd-EIT imaging. It is noted that different basis functions have significant impact on the quality of the fused images, especially for the radish. By employing the "coif3" or "dmey" basis function in the wavelet transform, the shapes of the nylon rob in the fused images are obviously distorted besides the shapes of the radish. Overall, the "fk14", "bior3.7", "sym6" and "db8" basis functions can provide fused images of reasonable quality, which are much better than those reconstructed by either td-EIT or fd-EIT.

4. Investigation for Improving Lung Ventilation Imaging with Thorax Model

To investigate the performance of the proposed method in medical EIT imaging of lung ventilation, simulations and in vitro experiments were conducted. A real-shaped thorax model, as shown in Figure 9, was built for FEM simulation. The thorax domain and the sub-domains of the lungs were created according to the build-in function of the thorax model in EIDORS [31]. It is well known that the whole respiratory cycle contains two stages: inspiration and expiration. In the simulation, the lung was simulated as varying states by setting different air contents in the given 2-D lung model. The model is assumed ideal as it does not consider the modelling errors regarding the non-homogenous background, the lung shape movement and the change of the chest due to the breathing in the real clinical situations. The following two test cases were designed:

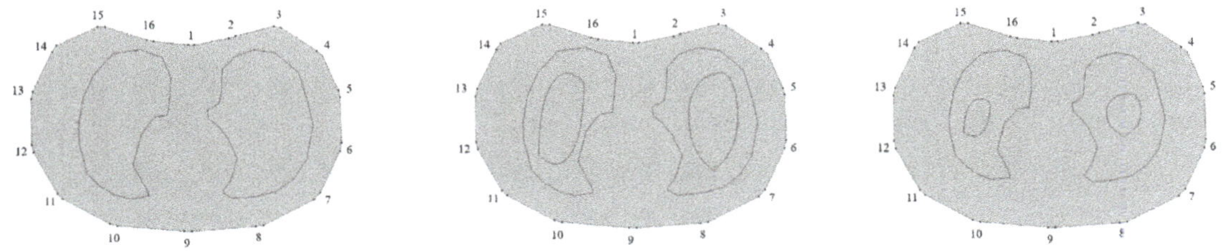

(**a**) Thorax model with lung (**b**) Inspiration model (**c**) Expiration model

Figure 9. 2-D view of thorax model built in simulation. (**a**) Thorax model with lung, (**b**) inspiration model, and (**c**) expiration model.

Case 1: Lung imaging in the inspiration (Figure 9b) stage, during which the breathing would increase the air content, and air would almost fully fill the lung in the end of inspiration. For this reason, it is suggested to set up a sufficiently large amount of air content inside the lung model to simulate the inspiration.

Case 2: Lung imaging in the expiration (Figure 9c) stage, during which the air content decreases with the breathing and becomes negligible in the end of expiration. In a similar fashion, a small amount of air content was set up inside the lung model to simulate the expiration.

In the simulation, lung conductivity was considered as a function of the excitation frequency, according to literature [32,33]. The maximum and minimum conductivity values were set to 0.079 S/m and 0.11 S/m, respectively. The conductivity of background tissues was set to 0.2 S/m, as suggested by Liu et al. [6]. Sixteen electrodes were attached on the boundary of the thorax domain. The same current injection and measurement strategy employed in the previous experimental studies was also adopted in the simulation. The simulation was conducted with 2-D FEM approximation in COMSOL. To produce a reference for fd- and td-EIT imaging, the reference frequency was chosen to be 1 kHz, and the reference distribution for td-EIT imaging was assumed to be the thorax model only containing the background medium and lung tissues, as shown in Figure 9a.

Simulation results are shown in Figures 10 and 11. In the reconstructed images, whose size is 84 $\times$ 128 pixels, the outermost dark solid line and the innermost solid dark line denote the true lung shape and air region, respectively. In the simulated cases 1 and 2, the fd- and td-EIT imaging can successfully reconstruct the lung-shaped inclusion or the air inclusion, respectively, but not both of them. In addition, the air inclusion is overestimated by td-EIT, while the lung-shaped inclusion is underestimated by fd-EIT. In fact, the end of expiration would reduce the content of air in the lung, and hence result in a higher RCR (the air domain is significantly overestimated) in the case 2, as shown in Tables 4 and 5. Due to the limitation of the reconstruction algorithm, the sharp edge and accurate shape of the inclusions were not well preserved in the reconstructed images. Note that the fd-EIT images at 1 kHz cannot be obtained since it is the chosen reference frequency.

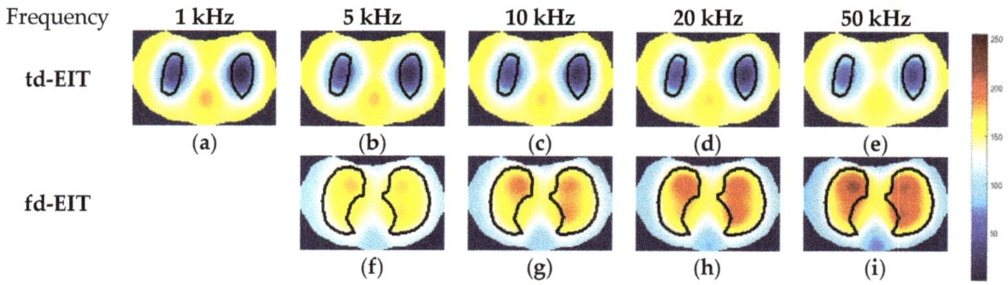

Figure 10. Td- and fd-EIT imaging results in case 1 with simulated lung. **(a–e)** are td-EIT images in case 1 at 1, 5, 10, 20 and 50 kHz, respectively; **(f–i)** are fd-EIT images in case 1 at 5, 10, 20 and 50 kHz, respectively.

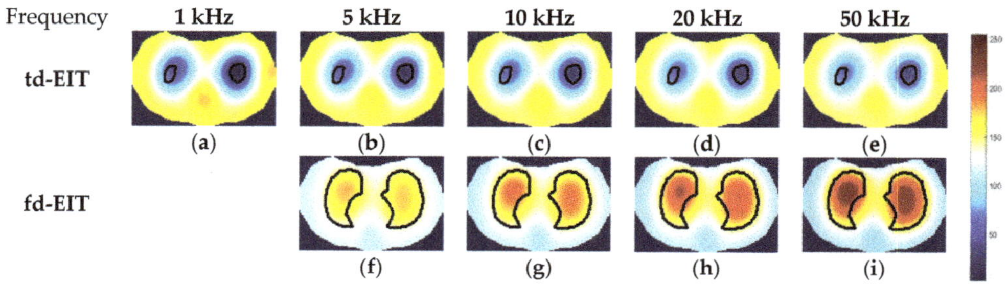

Figure 11. Td- and fd-EIT imaging results in case 2 with simulated lung. **(a–e)** are td-EIT images in case 2 at 1, 5, 10, 20 and 50 kHz, respectively; **(f–i)** are fd-EIT images in case 2 at 5, 10, 20 and 50 kHz, respectively.

Table 4. RCR and RIE regarding td- and fd-EIT imaging results in case 1.

Frequency		1 kHz	5 kHz	10 kHz	20 kHz	50 kHz
RCR	fd-lung	\	0.9063	0.9137	0.9207	0.9470
	td-air	1.2380	1.1077	1.0575	1.0010	0.9381
RIE	fd-EIT	\	0.4030	0.4032	0.4034	0.4041
	td-EIT	0.2395	0.2354	0.2339	0.2325	0.2308

Table 5. RCR and RIE regarding td- and fd-EIT imaging results in case 2.

Frequency		1 kHz	5 kHz	10 kHz	20 kHz	50 kHz
RCR	fd-lung	\	0.9064	0.9145	0.9237	0.9446
	td-air	4.1290	3.3334	2.9910	2.6587	2.2256
RIE	fd-EIT	\	0.4169	0.4168	0.4169	0.4172
	td-EIT	0.2232	0.2214	0.2208	0.2204	0.2200

The reconstructed images by td- and fd-EIT at 5 kHz and 50 kHz were used to demonstrate the improvement of the lung imaging by the proposed fusion method. As shown in Figures 12 and 13, the image fusion for lung imaging was fairly successful. Compared with the imaging results by either td-EIT or fd-EIT, the fused images cannot only provide the rational changes of air content, but also the position and shape of the lung-shaped inclusion. It is not trivial in practice to obtain the change of air content and lung shape information simultaneously since the latter can serve as a reference and there may, in addition to the air current, be tissue changes related to certain diseases such as pulmonary edema. Meanwhile, the size and shape of the air inclusion or lung-shaped inclusion in the fused image agree well with the actual ones. It is noted that the fused images at 50 kHz indicate a much larger image contrast in terms of the lung shape. This is in accordance with the finding that the conductivity variation of the lung tissues between the measurement frequency (5 kHz and 50 kHz) and the reference frequency (1 kHz) increases with frequency.

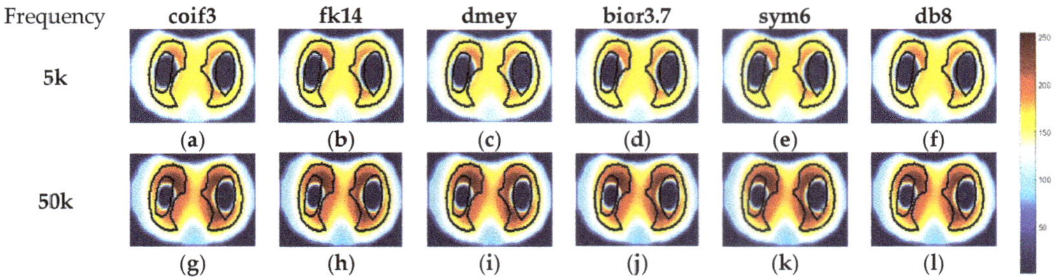

Figure 12. Fused images by applying wavelet-based fusion method in case 1. (**a**–**f**) are fused images using different wavelet bases in case 1 at 5 kHz, respectively; (**g**–**l**) are fused images using different wavelet bases in case 1 at 50 kHz, respectively.

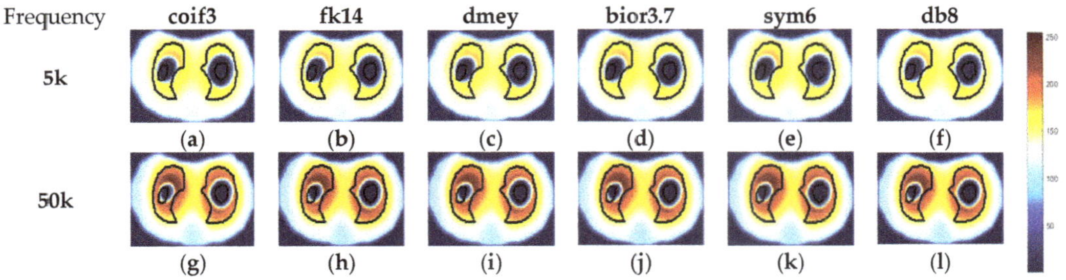

Figure 13. Fused images by applying wavelet-based fusion method in case 2. (**a**–**f**) are fused images using different wavelet bases in case 2 at 5 kHz, respectively; (**g**–**l**) are fused images using different wavelet bases in case 2 at 50 kHz, respectively.

To further validate the feasibility and effectiveness of the proposed fusion method, static in vitro experiments were conducted. These experiment was conducted in the same tank as in the previous experiments. The tank was also filled with tap water as the background medium, and two lung-shaped inclusions were made of white gourd. Two pairs of air-containing inclusions were inserted into the two lung-shaped inclusions to simulate the lung-filling conditions in the expiration and inspiration stages, respectively, as in the two simulation cases. Figures 14 and 15(a1,b1) show the phantom pictures. To quantify the air content inside the lung-shaped inclusion, the volume ratio was measured. The volume ratio was 3/20 and 6/20 for the expiration and inspiration setup, respectively.

Based on the visual inspection of Figures 14 and 15, the fused images provide a reasonable estimation of the shape and location of the lung-shaped inclusion as well as the air inclusion inside it at different frequencies. The td-EIT performs well in reconstructing the air inclusion, while the fd-EIT is better in imaging the lung-shaped inclusion. It is noted that the contrast variation caused by the change of air content is clearly shown. Therefore, it can be concluded that the lung imaging based on the proposed fusion method is feasible and effective.

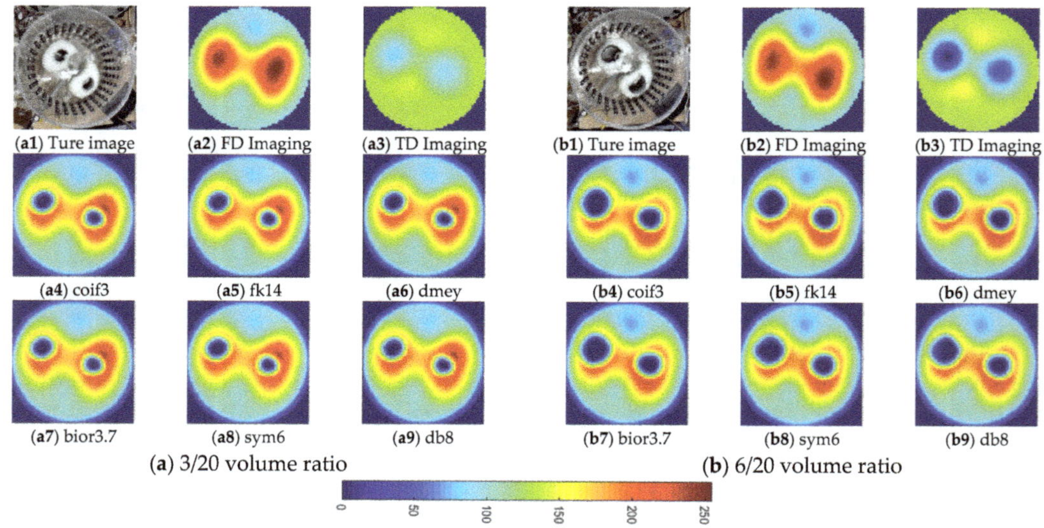

Figure 14. Reconstructed and fused images of lung-shaped white gourd containing a varied content of air at 10 kHz. (a) 3/20 volume ratio: (**a1**) is the actual experimental image; (**a2**,**a3**) are fd-EIT and td-EIT images, respectively; (**a4**–**a9**) are fused images using different wavelet bases, respectively. (**b**) 6/20 volume ratio: (**b1**) is the actual experimental image; (**b2**,**b3**) are fd-EIT and td-EIT images, respectively; (**b4**–**b9**) are fused images using different wavelet bases, respectively.

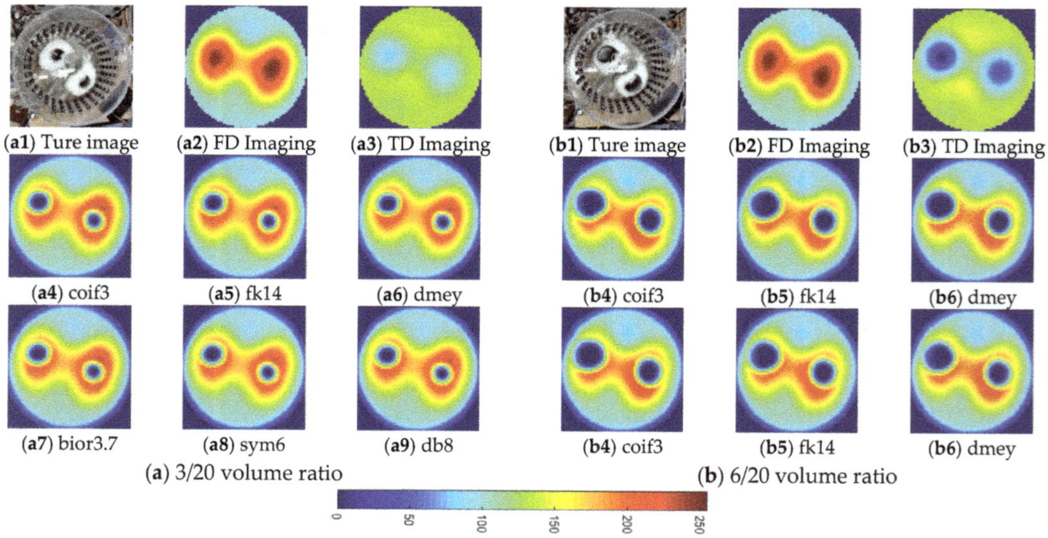

Figure 15. Reconstructed and fused images of lung-shaped white gourd containing a varied content of air at 20 kHz. (**a**) 3/20 volume ratio: (**a1**) is the actual experimental image; (**a2**,**a3**) are fd-EIT and td-EIT images, respectively; (**a4**–**a9**) are fused images using different wavelet bases, respectively. (**b**) 6/20 volume ratio: (**b1**) is the actual experimental image; (**b2**,**b3**) are fd-EIT and td-EIT images, respectively; (**b4**–**b9**) are fused images using different wavelet bases, respectively.

5. Conclusions

This paper characterizes and compares fd- and td-EIT for the imaging of target distributions composed of bio-material inclusions and non-conductive inclusions, and the lung ventilation imaging is a potentially applicable scenario. Through an experimental study with a peeled radish or a nylon rod as targets, it was found that td-EIT reconstructions can reconstruct the bio-material inclusion (the radish) and the non-conductive inclusion (the nylon rod) if either the former or the latter exists in the sensing domain. Fd-EIT could only image the bio-materials inclusion (the radish) as the conductivity of non-conductive materials is not frequency-dependent, but has significant distortions with an appropriate frequency interval. However, both the td-EIT and fd-EIT failed to simultaneously reconstruct the radish and nylon rod if both of them exist in the sensing domain. This is because the response to the non-conductive rod (i.e., a much larger conductivity contrast to the background medium) would overwhelm that to the radish for td-EIT imaging, which would occur when imaging certain lung diseases such as pulmonary edema. To tackle with this dilemma, it is proposed to fuse the td- and fd-EIT imaging results using a dedicated wavelet-based fusion strategy, which can integrate the merits of td- and fd-EIT. Based on the simulation and experiment with lung-shaped inclusions or in vitro lung models, it was found that the proposed method has much better reconstruction accuracy than either td- or fd-EIT, which can simultaneously reconstruct both the lung-shaped inclusions and air inclusions in the fused image. Meanwhile, the proposed method can reduce the image distortions by td- or fd-EIT, and provide the relative shapes, sizes, and positions of those inclusions. In summary, the proposed method can improve the performance of EIT in lung ventilation imaging, so that the diagnosis functions of EIT can be expanded to deal with more lung-tissue-related diseases. Although such a fd- and td-EIT system certainly requires the hardware to accommodate to both the EIT modes, which can be readily achieved in practical applications, this paper aims to determine how much improvement can be expected by applying such a fusion method and understanding and identifying the compromises involved in fd- and td-EIT imaging. Future work will focus on optimizing the image reconstruction and fusion methods to further enhance the image quality. On other hand, the proposed method should be tested in more realistic cases of lung imaging.

Author Contributions: Methodology, formal analysis, software, and validation, X.B. (Xue Bai); methodology, original draft preparation, review and editing, D.L. and W.T.; methodology and formal analysis, J.W. and X.B. (Xu Bai); funding acquisition, review and editing, S.S. All authors have read and agreed to the published version of the manuscript.

Funding: This research was funded by Beijing Advanced Innovation Center for Biomedical Engineering.

Institutional Review Board Statement: Not applicable.

Informed Consent Statement: Not applicable.

Data Availability Statement: Not applicable.

Conflicts of Interest: The authors declare no conflict of interest.

References

1. Brown, B.H. Electrical impedance tomography (EIT): A review. *J. Med. Eng. Technol.* **2003**, *27*, 97–108. [CrossRef] [PubMed]
2. Adler, A.; Arnold, J.H.; Bayford, R.; Borsic, A.; Brown, B.; Dixon, P.; Faes, T.; Frerichs, I.; Gagnon, H.; Garber, Y. GREIT: A unified approach to 2D linear EIT reconstruction of lung images. *Physiol. Meas.* **2009**, *30*, S35–S55. [CrossRef] [PubMed]
3. Yang, Y.; Jia, J. A multi-frequency electrical impedance tomography system for real-time 2D and 3D imaging. *Rev. Sci. Instrum.* **2017**, *88*, 085110. [CrossRef] [PubMed]
4. Lee, K.; Woo, E.J.; Seo, J.K. A Fidelity-embedded Regularization Method for Robust Electrical Impedance Tomography. *IEEE Trans. Med. Imaging* **2017**, *37*, 1970–1977. [CrossRef] [PubMed]
5. Liu, D.; Smyl, D.; Du, J. A Parametric Level Set-Based Approach to Difference Imaging in Electrical Impedance Tomography. *IEEE Trans. Med. Imaging* **2019**, *38*, 145–155. [CrossRef] [PubMed]
6. Liu, D.; Smyl, D.; Du, J. Nonstationary Shape Estimation in Electrical Impedance Tomography Using a Parametric Level Set-Based Extended Kalman Filter Approach. *IEEE Trans. Instrum. Meas.* **2020**, *69*, 1894–1907. [CrossRef]
7. Henderson, R.P.; Webster, J.G. An Impedance Camera for Spatially Specific Measurements of the Thorax. *IEEE Trans. Biomed. Eng.* **1978**, *25*, 250–254. [CrossRef]
8. Adler, A.; Boyle, A. Electrical Impedance Tomography: Tissue Properties to Image Measures. *IEEE Trans. Biomed. Eng.* **2017**, *64*, 2494–2504.

9. Mellenthin, M.M.; Mueller, J.L.; Camargo, E.D.; Moura, F.D.; Santos, T.; Lima, R.G.; Hamilton, S.J.; Muller, P.A.; Alsaker, M. The ACE1 Electrical Impedance Tomography System for Thoracic Imaging. *IEEE Trans. Instrum. Meas.* **2018**, *68*, 3137–3150. [CrossRef]
10. Barber, D.C.; Brown, B.H. Applied potential tomography. *J. Br. Interplanet. Soc.* **1989**, *42*, 391. [CrossRef]
11. Seo, J.K.; Harrach, B.; Woo, E.J.; Ammari, H. Recent progress on Frequency Difference Electrical Impedance Tomography. *ESAIM Proc.* **2009**, *26*, 150–161. [CrossRef]
12. Seo, J.K.; Lee, J.; Kim, S.W.; Zribi, H.; Woo, E.J. Frequency-difference electrical impedance tomography (fdEIT): Algorithm development and feasibility study. *Physiol. Meas.* **2008**, *29*, 929–944. [CrossRef]
13. Packham, B.; Koo, H.; Romsauerova, A.; Ahn, S.; Mcewan, A.; Jun, S.C.; Holder, D.S. Comparison of frequency difference reconstruction algorithms for the detection of acute stroke using EIT in a realistic head-shaped tank. *Physiol. Meas.* **2012**, *33*, 767. [CrossRef]
14. Wilson, A.J.; Milnes, P.; Waterworth, A.R.; Smallwood, R.H.; Brown, B.H. Mk3.5: A modular, multi-frequency successor to the Mk3a EIS/EIT system. *Physiol. Meas.* **2001**, *22*, 49–54. [CrossRef]
15. Mcewan, A.; Romsauerova, A.; Yerworth, R.; Horesh, L.; Bayford, R.; Holder, D. Design and calibration of a compact multi-frequency EIT system for acute stroke imaging. *Physiol. Meas.* **2006**, *27*, S199. [CrossRef] [PubMed]
16. Tong, I.O.; Koo, H.; Lee, K.H.; Sang, M.K.; Woo, E.J. Validation of a multi-frequency electrical impedance tomography (mfEIT) system KHU Mark1: Impedance spectroscopy and time-difference imaging. *Physiol. Meas.* **2008**, *29*, 295–307.
17. Oh, T.I.; Wi, H.; Kim, D.Y.; Yoo, P.J.; Woo, E.J. A fully parallel multi-frequency EIT system with flexible electrode configuration: KHU Mark2. *Physiol. Meas.* **2011**, *32*, 835–849. [CrossRef]
18. Zhang, K.; Guo, R.; Li, M.; Yang, F.; Abubakar, A. Supervised Descent Learning for Thoracic Electrical Impedance Tomography. *IEEE Trans. Biomed. Eng.* **2020**, *68*, 1360–1369. [CrossRef] [PubMed]
19. Santos, S.A.; Czaplik, M.; Orschulik, J.; Hochhausen, N.; Leonhardt, S. Lung Pathologies analyzed with Multi-Frequency Electrical Impedance Tomography: Pilot Animal Study. *Respir. Physiol. Neurobiol.* **2018**, *254*, 1–9. [CrossRef]
20. Li, S.; Kang, X.; Fang, L.; Hu, J.; Yin, H. Pixel-level image fusion: A survey of the state of the art. *Inf. Fusion* **2017**, *33*, 100–112. [CrossRef]
21. Hua, P.; Woo, E.J. Iterative reconstruction methods using regularization and optimal current patterns in electrical impedance tomography. *IEEE Trans. Med. Imaging* **1991**, *10*, 621–628. [CrossRef]
22. Polydorides, N.; Lionheart, W. A Matlab toolkit for three-dimensional electrical impedance tomography: A contribution to the Electrical Impedance and Diffuse Optical Reconstruction Software project. *Meas. Sci. Technol.* **2002**, *13*, 1871–1883. [CrossRef]
23. Lionheart, W.R. EIT reconstruction algorithms: Pitfalls, challenges and recent developments. *Physiol. Meas.* **2004**, *25*, 125–142. [CrossRef]
24. Peijia, Z.; Jiwu, H. Discrete Wavelet Transform and Data Expansion Reduction in Homomorphic Encrypted Domain. *IEEE Trans. Image Process.* **2013**, *22*, 2455–2468. [CrossRef] [PubMed]
25. Lei, Z.; Jiangbao, Y.; Feng, Z.; Xiangyu, T.; Lifeng, Z. Reconstruction Method of Electrical Capacitance Tomography Based on Wavelet Fusion. In *MATEC Web of Conferences, Proceedings of the 2018 6th International Forum on Industrial Design (IFID 2018), Luoyang, China, 18–20 May 2018*; EDP Sciences: Les Ulis, France, 2018; p. 176.
26. Li, L.; Zhang, Y.; Song, L. An Image Fusion Algorithm for ECT based on Tikhonov Algorithm and Wavelet Transform. *Int. J. Signal. Process. Image Process. Pattern Recognit.* **2014**, *7*, 51–60. [CrossRef]
27. Yu, W.; Wang, M.; Zhang, Z. Study on Registration and Fusion Multimodality Medical Image. *J. Converg. Inf. Technol.* **2013**, *8*, 163–169.
28. Tian, W.; Suo, P.; Liu, D.; Sun, S.; Xu, L. Simultaneous Shape and Permittivity Reconstruction in ECT with Sparse Representation: Two-Phase Distribution Imaging. *IEEE Trans. Instrum. Meas.* **2021**, *70*, 1–14. [CrossRef]
29. Bera, T.K.; Nagaraju, J.; Lubineau, G. Electrical impedance spectroscopy (EIS)-based evaluation of biological tissue phantoms to study multifrequency electrical impedance tomography (Mf-EIT) systems. *J. Vis.* **2016**, *19*, 1–23. [CrossRef]
30. Li, J.; Wan, N.; Wen, J.; Cheng, G.; Cheng, L. Quantitative detection and evaluation of thrombus formation based on electrical impedance spectroscopy. *Biosens. Bioelectron.* **2019**, *141*, 111437. [CrossRef] [PubMed]
31. Adler, A.; Lionheart, W. Uses and abuses of EIDORS: An extensible software base for EIT. *Physiol. Meas.* **2006**, *27*, S25–S42. [CrossRef]
32. Gabriel, S.; Lau, R.W.; Gabriel, C. The dielectric properties of biological tissue II: Measurements in the frequency range 10 Hz to 20 GHz. *Phys. Med. Biol.* **1996**, *41*, 2251–2269. [CrossRef] [PubMed]
33. Gabriel, S.; Lau, R.W.; Gabriel, C. The dielectric properties of biological tissues: III. Parametric models for the dielectric spectrum of tissues. *Phys. Med. Biol.* **1996**, *41*, 2271–2293. [CrossRef] [PubMed]

biosensors

MDPI

Review

EMG-Centered Multisensory Based Technologies for Pattern Recognition in Rehabilitation: State of the Art and Challenges

Chaoming Fang [1,†], Bowei He [2,†], Yixuan Wang [1], Jin Cao [3] and Shuo Gao [1,4,*]

1 School of Instrumentation and Optoelectronic Engineering, Beihang University, Beijing 100083, China; timefliesfang@buaa.edu.cn (C.F.); wangyixuan@buaa.edu.cn (Y.W.)

2 School of Automation Science and Electrical Engineering, Beihang University, Beijing 100083, China; 17374556@buaa.edu.cn

3 Department of Psychiatry, Massachusetts General Hospital, Harvard Medical School, Boston, MA 02138, USA; jcao9@mgh.harvard.edu

4 Beijing Advanced Innovation Center for Big Data-Based Precision Medicine, Beihang University, Beijing 100083, China

* Correspondence: shuo_gao@buaa.edu.cn

† These authors contribute equally to this work.

Received: 27 June 2020; Accepted: 22 July 2020; Published: 26 July 2020

Abstract: In the field of rehabilitation, the electromyography (EMG) signal plays an important role in interpreting patients' intentions and physical conditions. Nevertheless, utilizing merely the EMG signal suffers from difficulty in recognizing slight body movements, and the detection accuracy is strongly influenced by environmental factors. To address the above issues, multisensory integration-based EMG pattern recognition (PR) techniques have been developed in recent years, and fruitful results have been demonstrated in diverse rehabilitation scenarios, such as achieving high locomotion detection and prosthesis control accuracy. Owing to the importance and rapid development of the EMG centered multisensory fusion technologies in rehabilitation, this paper reviews both theories and applications in this emerging field. The principle of EMG signal generation and the current pattern recognition process are explained in detail, including signal preprocessing, feature extraction, classification algorithms, etc. Mechanisms of collaborations between two important multisensory fusion strategies (kinetic and kinematics) and EMG information are thoroughly explained; corresponding applications are studied, and the pros and cons are discussed. Finally, the main challenges in EMG centered multisensory pattern recognition are discussed, and a future research direction of this area is prospected.

Keywords: multisensory; electromyography; pattern recognition; rehabilitation

1. Introduction

Monitoring and analyzing the patient's physiological information are of significance in the process of physical rehabilitation in order to evaluate the rehabilitation effect [1,2] and control auxiliary devices during the physical rehabilitation process. Conventionally, the physiological information is divided into physical and psychological information, e.g., muscle force information and the intention of the patient. To detect these two categories of information, various types of sensors, including electromechanical sensors (such as accelerometers [3,4], gyroscopes [5,6] and force sensors [7,8]), and biosensors (such as electromyography (EMG) [9–11] magnetoencephalography (MEG) and electroencephalogram (EEG)) have been utilized. Electromechanical sensors are capable of detecting the physical information effectively. For example, in [12], flex sensors and force sensitive sensors (FSRs) are utilized to detect the bending angle of the finger and the grasp force, respectively. The sensing information is then used for the control of robotic fingers in performing ten different tasks. In [13], the integrated use of accelerometer sensors, pulse sensors, and ambient sensors is reported for recognizing sedentary behavior with an accuracy of 95%. Nevertheless, biosensors are not only proven to reflect physical information, but convert the bioelectric signal, which is broadly treated as a direct link to human psychological information, into interpretable voltage amplitudes, and therefore have unique advantages in psychological detection over the electromechanical sensors [11,14]. Note that in the rehabilitation area, human psychological information normally refers to human intention information, which can be used to examine the synchronization level between human movements and human thinking. Hence, in this article, the detection of human intention, instead of human psychology, is focused.

Among all biosensor-captured information, EEG, MEG, and EMG are the three most relevant signals to human intention [15–17]. Among them, the EEG signal is of weak robustness due to the shortage of noninvasive electrodes in collecting a surface EEG signal, failing to provide high signal quality. MEG techniques, e.g., MRI, can offer accurate and rich information, but the time-consuming issue and huge machine volume block its use in rehabilitation. In contrast, EMG techniques enjoy relatively higher signal-noise ratio (SNR) and robustness than EEG means (especially during movements), and process information much faster than MEG techniques. Therefore, EMG is a more preferred choice for intention detection in the rehabilitation field.

Two widely used techniques to interpret the EMG signal are signal intensity registration and pattern recognition [18,19]. The former approach is based on the correlation between EMG signal intensity and quantified muscle force level [20,21]. However, the EMG amplitude of a muscle is usually determined by many uncontrollable factors, e.g., EMG crosstalk [22] and variation in muscle force [23], which will not only generate misinterpretations of the muscle activities, but also result in potential safety risks when such information is directly utilized in the exoskeleton for the motion functions impaired patients [18]. Alternatively, the latter approach offers higher accuracy in decoding EMG signals owing to the higher level of information, e.g., a set of motions or the movement intention can be extracted via raw data to provide deep insights in examining user's body condition. EMG PR was not popularized for its high computational cost compared to its counterpart, until being promoted by the rapid development of electronics and information technology in recent years. Now EMG PR has been broadly applied to disease diagnosis [24], intelligent prosthesis control [19], and gains continuing attentions from relevant researchers.

Generally, the recognition scenarios can be divided into upper limb PR and lower limb PR. In early years, research on EMG PR mainly focused on the recognition of the upper limb movements with great differences, e.g., largely changed arm movements [25]. This is because the EMG patterns of such movements are significantly different and therefore is easier to be recognized. Nevertheless, fine upper movements and lower limb movements are also important in many neuromuscular disease analysis, such as Parkinson's disease and myasthenia. Therefore, it is also important to create a reliable relationship between EMG signal and patients' intentions and physical conditions for these two kinds of movements.

However, using the EMG signal to interpret fine upper limb and lower limb motions is challenging. In terms of the former, small movements in the upper limbs like the finger or wrist only result in slight differences in EMG signals, giving rise to difficulties in effectively distinguishing the EMG signal of one pattern from others. For example, a hand movement recognition method using single-channel sEMG is presented in [26]. This work reached an accuracy of 86.7% in classifying nine finger movements, and the recognition accuracy and the pattern types are all lower than the multisensory approach [27]. As to the latter, interactions between lower limbs and unpredictable environments during movements can generate unexpected EMG patterns. For example, the muscle activity of lower limbs is often related to the flatness of terrains [28], or the different locomotion modes such as obstacle crossing [29] and stairs ascending/descending [30]. Therefore, utilizing only the EMG signal cannot satisfy the demand for reaching robust recognition performances at these complicated scenarios without learning environmental knowledge. A possible solution is to concurrently acquire the external information of the environment as well as the EMG signals, so that information from different dimensions can be integrated for further analysis and complementing each other.

In light of this, EMG-centered multisensory-based technologies emerge in recent years, and increasing demonstrations and attempts in three main rehabilitation scenarios have been reported globally. First, multisensory fusion EMG PR plays an important role in accurate and early diagnosis for neuromuscular diseases so that the patients could start the rehabilitation process before the diseases further develop. For example, in [31], the acceleration information of the muscles and forearm of the biceps brachii of the Parkinson's patient are collected simultaneously, 12 kinds of signal features were extracted for cluster analysis, and successfully achieved a diagnostic accuracy rate of more than 90%. Second, multi-sensor fusion can evaluate the rehabilitation process of patients and the effectiveness of the treatment in more dimensions. For example, in [32], the fusion analysis of the plantar force information of the Parkinson's patient and the myoelectric information of the anterior tibial muscle. The effect of a plantar sensory stimulation therapy on improving gait and motor output in Parkinson's patients were evaluated. Third, multisensory EMG PR is used as the control strategy for assistive devices such as intelligent prosthesis and exoskeleton [33–36], these auxiliary devices can effectively help patients in physical rehabilitation, and to a certain extent reduce the burden of the physical therapy. In addition, multi-sensor fusion also allows these devices to obtain higher security and a more comfortable user experience. For example, in [37], an exoskeleton hand is designed to help paralyzed people in rehabilitation training. It uses myoelectricity as an input to represent the patient's intention and uses angle information and gripping force information as feedback to provide reference and correction for EMG signals in real-time, ensuring that the rehabilitation action can be completed accurately. The above three application scenarios are conceptually depicted in Figure 1.

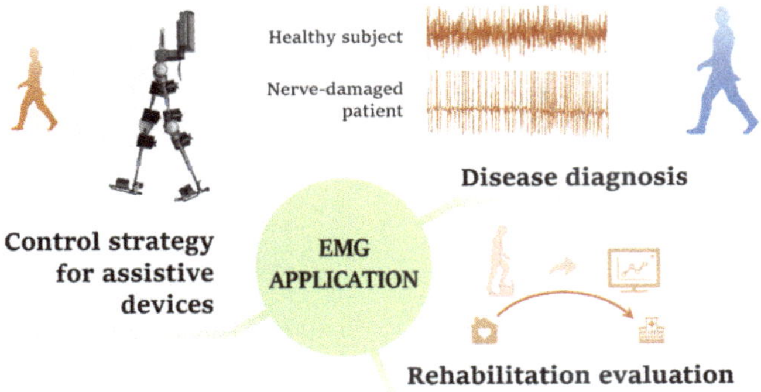

Figure 1. The conceptual depiction of the application scenarios of the electromyography (EMG) pattern recognition technique.

To provide a timely and systematic overview of the EMG based pattern recognition techniques using multisensory fusion strategies in the context of rehabilitation, this review article is composed. The paper is structured as follows: we first introduce the basic physiology knowledge relevant to EMG-based PR in Section 2. Then, in Section 3, a detailed explanation of the processing procedure and comparison of the state-of-the-art EMG PR is presented. Afterward, Section 4 explains the collaborations between EMG and multi-sensory information and presents diverse multisensory strategies successfully applied in different rehabilitation scenarios. Finally, the challenges faced by EMG PR and the prospect in resolving these challenges are summarized in Section 5.

2. Physiology Background

For reader convenience in understanding the working principle for EMG pattern recognition, some relevant physiological backgrounds, including the mechanism of EMG generation and the target patterns, are essential prior knowledge. This chapter will explain these physiological backgrounds, which helps readers to understand the characteristics of EMG signals and the design of recognition systems for the target patterns.

2.1. EMG Signal Overview

EMG is a type of technique used to evaluate and record a series of electrical signals that emanate from body muscles [38]. The principle of the generation of the EMG signal can be described as follows. Motor nerve cells produce electrical pulses under the control of the central nervous system in the cerebral cortex. These neural signals are transmitted to muscle fibers through axons and cause pulse sequences, which activate them to contract and produce muscle tension. Meanwhile, a current is generated in the human body that brings about transmembrane potential [39]. The transmembrane potential is the difference between the internal and external potentials of muscle cell membranes. When muscle cells are in a quiet state, the cell membrane potential is polarized. The potential difference between the inside and outside of the cell membrane under the polarization state is called resetting potential. Depolarization occurs when the cell is excited, and this trend will spread to around [40]. The corresponding action potential is defined as an electromyography signal. The above procedure is demonstrated in Figure 2.

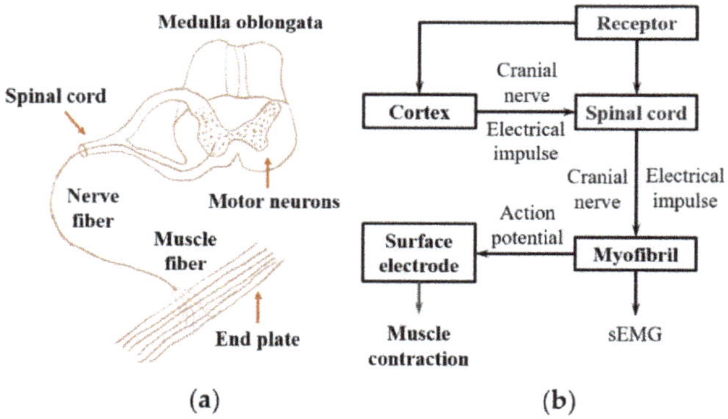

Figure 2. The principle of generation of the EMG signal. (**a**) the structure of the neuro-muscular system. (**b**) the schematic of the EMG signal transduction in the nerve and muscle system.

2.2. Human Movement Patterns

As the EMG signal is a bioelectrical signal controlled by the mutual effect of the receptors and nerve system of human beings, the pattern of the EMG signal relies on both the user's subjective intention and the interactive environment conditions while performing specific actions. In this section, we will introduce some frequently studied upper and lower limb patterns in relevant studies.

The upper limbs of the human body include a forearm, elbow, rear arm, wrist, and hand. The muscle mass of the upper limb is generally small and slender, and the movement range of the upper limb is not extensive. Compared with the lower limbs, the muscle strength of the upper limbs is generally weaker because the body does not need to be supported.

The main functions of the upper limbs are grasping, stretching, and expressing information with hand gestures. Therefore, the patterns in the upper limb can be divided into two categories: limb position and hand gestures. The classification of Limb positions has been elaborated comprehensively in [25], including P1: arm hanging at the side, elbow bent at 90 degrees; P2: straight arm reaching up (45 degrees from vertical); P3: straight arm hanging at side; P4: straight arm reaching forward. Hand gestures account for the dexterous movements of the hand, including wrist and finger movements. Wrist gestures are the hand movements that rotate the whole hand around the wrist joint, but finger gestures only involve the movements of the fingers. Reference [27] proposed a challenging set of 15 gestures in which 5 (thumb, index, middle, ring, pinky) are finger gestures, 6 are wrist gestures. The above patterns are demonstrated in Figure 3.

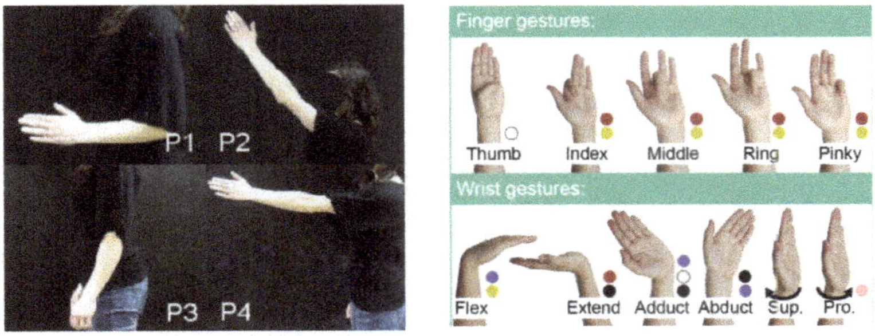

(a) (b)

Figure 3. Two types of upper limb patterns. (a) the limb position, adopted from [25]. (b) hand gestures, adopted from [27].

The lower limbs of the human body include the hip, thigh, calf, foot, hip, knee, ankle, and other joint parts. The lower limb has the characteristics of larger muscle tissue and a larger joint movement angle [41]. Its function is mainly to support the human body and change the position of the human body. Depending on these physiological functions of the lower limbs, the activities most closely related to the lower limbs are often activities with the core of walking behavior, so the activity recognition based on the lower limbs can be roughly summarized in two aspects, namely, the locomotion mode and the gait phase.

The first one is locomotion mode, which aims at the process that the brain actively coordinates the limbs to make different actions when the human body is facing different environments in the real world. Specifically, up and downslope, walking on flat ground, up and downstairs are the most widely studied in the relevant research [35,42,43]. This is not only because these five sports modes are the most common and widespread scenes in life, but also because these modes have the process of overcoming the influence of gravity and changing the center of gravity of the body, so auxiliary equipment is needed to adjust the lower limbs [44,45]. In addition to these five kinds of sports modes, some work has also been done to study such sports modes as crossing obstacles [29,46], turning [29], standing, and sitting transition [47]. These locomotion modes are a more in-depth interpretation than simple actions, such as leg swinging, touchdown, etc., which is conducive to the auxiliary equipment to have a better understanding of the user's environment and generate adaptive control based on this.

Gait phase is another widely concerned lower limb mode, which is a scientific decomposition and utilization of human periodic walking movement. There are different ways to define the gait phases according to different usage scenarios. The simplest one is to divide the gait into two phases: namely, the stance phase and swing phase [3]. Generally, the heel strike (HC) and toe-off (TO) are used as the starting point of the stance phase and the swing phase, respectively [48]. However, this division is not accurate enough and is not applicable in scenes that require high continuity. In [42], a definition of the gait phase consisting of four phases is proposed and used for real-time recognition. In the most precise definitions, the gait cycle can be divided into eight phases [49].

The function of the gait phase is to provide auxiliary equipment to refine the human walking process and provide different auxiliary control in different gait segments [42,50]. The characteristics of the gait phase, such as continuity, duration, and so on, to describe the whole walking cycle of the human body, is a standard method in gait analysis and has been proved to be related to the diagnosis of some diseases [51]. Figure 4 shows the five most commonly studied locomotion modes, as well as a detailed definition of the gait phases.

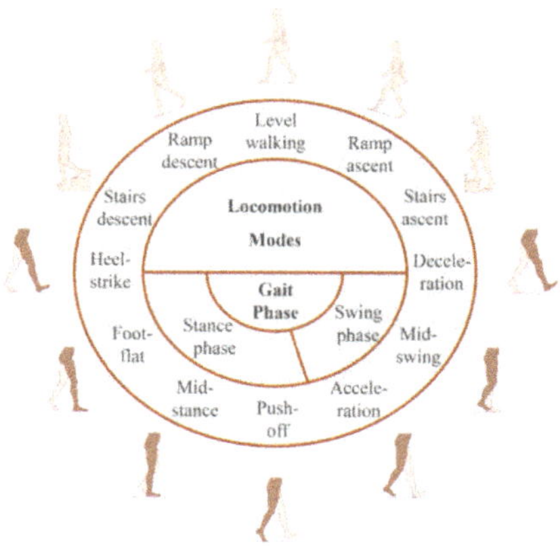

Figure 4. The two commonly studied lower-limb patterns, i.e., locomotion modes and gait phase. Five locomotion modes, along with the definition of the eight-phases gait cycle, are presented here.

3. EMG Pattern Recognition Pipeline

The differences in human body motion patterns can be reflected in EMG signals. However, the raw EMG data is usually very noisy and cannot be directly classified. Therefore, a series of processing procedures is necessary to distinguish these differences accurately. Generally, the PR procedure can be summarized in the following four steps (also depicted in Figure 5):

1. Data acquisition
2. Signal preprocessing
3. Feature extraction and reduction
4. Classification

There is no universal standard for these processing procedures yet, owing to the variability for EMG PR in different rehabilitation scenarios. Therefore, in this section, we will explain the EMG PR protocols and approaches adopted by relevant studies in detail.

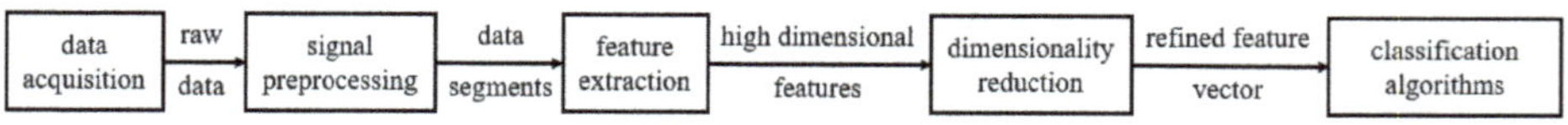

Figure 5. The pipeline of performing EMG pattern recognition.

3.1. Data Acquisition

Acquiring a high-quality EMG signal is strongly desired to ensure a satisfying recognition accuracy level, so properly designing the front-end EMG acquisition approaches is very important. In this section, we will compare the methods in front-end EMG acquisition from two aspects: the design of the EMG sensing system and the selection of muscle measurement points.

3.1.1. EMG Sensing System

The sensing system for collecting EMG signals is mainly composed of electrodes, amplifiers, microprocessors, and transmission devices. The electric signal generated by the muscle is picked up by the electrodes, amplified by the amplification circuit, and then transmitted to the host computer via the transmission device. Generally, the connection between electrodes and processor is wired while the transmission is typically wireless.

In the collection of EMG, the design of the sensor module is typically highly similar, except for the electrode. Common EMG electrodes include two main kinds of pole placement methods: monopolar electrode and bipolar electrode. Both placing methods measure the potential with reference to the electrode placed at the locations without EMG response (e.g., ankle or knee), while the Monopoles method directly measures the potential difference and bipolar method apply the differentiate amplification method [52]. Bipolar electrodes have a higher usage frequency because the common mode noise can be suppressed in real application but lack of setup flexibility when compared to monopoles. Besides, the EMG signal quality is also influenced by the distance between each electrode pole and their diameters and widths when bipolar electrodes are applied [38,52].

In addition to electrode placement, the implantation of the electrode is also reviewed here. Intramuscular EMG and surface EMG are two mainstream Electromyography signal acquisition schemes. Usually, intramuscular EMG signals are recorded using percutaneous fine wire electrodes or others made of similar materials. These electrodes need to be inserted into several muscle tissues using hypodermic needles [53]. Insertion locations are identified by palpation [54] and verified by electrical stimulation and EMG channel activity during corresponding test contractions. Different from intramuscular EMG signal collection, surface EMG (sEMG) electric potentials are acquired with electrodes placed on the skin just above the target muscle [52]. Electrodes occurring in previous experiments are usually made of silver. Typically, there are two ways for operators to contact the electrode with the skin. One is by using a silver-chloride gel to achieve wet contact while the other applies dry electrodes with microneedles to record EMG signals [43].

According to the literature research, surface EMG hosts a higher usage frequency compared with intramuscular EMG [52,55–57]. The main reasons for its wide application are regarded to be non-invasive and convenient. Modern surface EMG electrodes are stuck to the surface of muscle and avoid the danger of causing potential muscle injuries to tissue, which is considered to be the critical concern in intramuscular EMG. Besides, during the use of sEMG, operators can freely select measurement points on the surface of muscle tissue and reduce the time cost to do the preparation work like disinfecting the probe points with alcohol. Nevertheless, some bold attempts in the application of intramuscular EMG have also been provided recently [53,54]. Intramuscular EMG provides another signal source to detect the human body behavior and handles some difficulties related to sEMG-based control, such as the unstable contact of the electrode with the skin. Some other additional benefits of intramuscular EMG are given in [53], for example, the ability to record deep muscle signals with little EMG crosstalk. Although the experimental results in [53,54,58] all prove that intramuscular EMG could not bring with the reduction in classification error from sEMG for single classifiers while the significant decrease of classification error for parallel classifiers can be achieved. According to the practical efforts in [53,54], the parallel configuration for simultaneous control becomes more promising via the use of intramuscular EMG, which has long been an unsolved problem in sEMG studies.

3.1.2. Muscle Site Selection

In the thesis of movement recognition, reasonable muscle group selection for EMG signal obtainment is of considerable significance for robust recognition. Various muscle groups in both the lower and upper limbs are functioning and positioned differently, and Figure 6 demonstrates the anatomical structure of the muscles.

Real muscle selection needs to consider the previous experience and specific demand in the application scenario. For example, as for prosthetic control, which is mainly designed for the amputees, experimenters commonly choose the proximal muscle groups like thigh muscle group [29,35,42] and the upper arms [59] because the distal ones are considered to be amputated. While, as for the studies oriented for non-amputees subjects, the distal muscle groups such as the shank muscle tissues and forearm and wrist muscles are often selected as targeted muscles due to their lower inter-subject variability than the proximal [43,60,61]. Gao Shuo [43] chose a pair of distal antagonistic muscles, tibialis anterior (TA) and soleus (SL) to conduct the terrain identification experiment and got satisfying results. Besides, muscle characteristic is also an important aspect to be taken into account. For example, the proximal hip muscle groups (AM, GM, and PRF) are taken into consideration in [9] due to their sensitivity to different walking speeds. It has been proved in [61] that rectus femoris (RF) and Soleus (SL) are responsible for pulling the body forward at the stance phase of the gait and therefore studies related to the recognition of uphill/upstairs behaviors demanding to overcome the gravity are more likely to choose such muscles for EMG detection [62]. The selected criterion and the specific muscle groups selected by relevant studies have been summarized in Table 1.

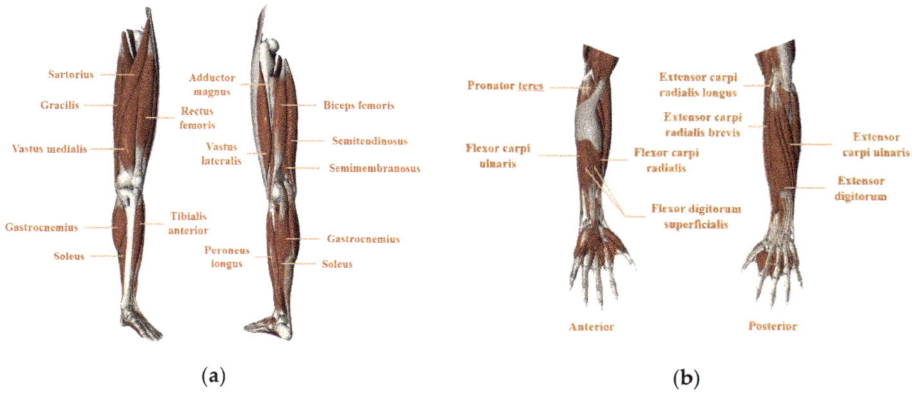

Figure 6. Longitudinal representation of the muscle groups of (**a**) the lower limb (shank and thigh). (**b**) the upper limb (elbow and hand).

3.2. Signal Preprocessing

The original EMG signals obtained through the front-end acquisition equipment are generally noisy, and pattern recognition requires the use of signals of limited length for further processing. Therefore, this section will introduce filtering and windowing preprocessing methods to reduce the noise in raw data and segmenting the EMG signals.

Real EMG signals are usually mixed with various kinds of noise due to environmental disturbance and other uncontrollable factors in the experiments. Reasonable preprocessing is necessary for extracting useful features in further analysis. Filtering, normalization, and windowing are the three most common EMG data preprocessing methods. The filtering technique derives from classical signal process schemes of signal preprocessing in the frequency domain. Most of the energy of the EEG signal are mainly within the frequency range 0–500 HZ [65]. In [66,67], it is stated that high-frequency (500–1000 HZ) EMG signal is likely to be interfered with by the aliasing. As for the low-frequency region (1–10 HZ), it is mainly regarded as a kind of noise caused by the cable movement and the interface of the measurement electrode and the skin. Actually, the most commonly used EMG signal filter is essentially a finite impulse response bandpass filter with a cutoff point of 10 HZ (low) and 500 HZ (high) [68–70].

Table 1. Summary of the selected muscle group for detecting electromyography (EMG) signals.

No.	Proximal Muscle	Distal Muscle	Comments
[42]	SAR, RF, VL, VM, GRA, BFL, SEM, BFS, ADM	/	Gluteal muscles (gluteus maximus and gluteus medius) on the amputated side and the thigh muscles of the residual limb were monitored
[35]	RF, VL, VM, BFL, SEM, BTS, TFL	/	The accurate electrodes locations are adjusted according to the able-bodied subjects and transfemoral subjects
[29]	SAR, RF, VL, VM, GRA, BFL, SEM, BFS, ADM	/	It should be noted that the locations of EMG electrodes on the distal muscles were approximate
[60]	/	TA, SL	Although only two muscles are selected, the classification accuracy is still satisfying
[59]	TPA, DPA, PMC, BCL, TBL, FCR, ECR	/	One of the eight signal channels is used for the synchronization of data from the Fastrack while the left seven are utilized to collect muscle activities signal.
[27]	/	FDS, FDP, EDC, EIP, EMP	These selected muscles are responsible for controlling all fingers except the thumb.
[9]	AM, GM, PRF, VL, VM	/	The proximal hip muscle groups have higher rates of the change in EMG activation with regard to different walking speeds while the distal knee extensor muscle groups show higher rates of change for different waling slopes
[61]	GM, RF, VL, BFL	TA, GA, SL	Humans often change gait patterns to prevent overexertion and possible injury to the relatively small dorsiflexor muscles, which are walking close to maximum capacity.
[63]	RF, VL, SEM		These three thigh muscles are the most commonly used muscles to classify locomotion modes at different speeds.
[64]	BF, RF	MG, TA	To reflect the effect of gait speed and gender on joint motion of lower extremity more comprehensively, bilateral lumbar erectors spinae are also utilized besides the muscles mentioned before.

Muscles: TA = Tibialis Anterior; SL = Soleus; SAR = Sartorius; RF = rectus femoris; VL = vastus lateralis, VM = vastus medialis; GRA = gracilis; BFL = biceps femoris long head; SEM = semitendinosus; BFS = biceps femoris short head; ADM = adductor magnus.

Due to the difficulty when dealing with the long-term sensor data sequence, windowing operation is necessary to slice it into short clips for the real-time prediction task. Generally, the properties of the signal decide the window length when analyzing the EMG data. According to [71], EMG signal can be considered a wide-sense Gaussian random process. Therefore, proper window length is of great significance for information extraction. The too-long window will lead to clips with an unbearable variance, while too shot sequence may not be able to contain enough useful information for classification [72]. Huang, He et al. [29] state that the analysis window length should not exceed 200 ms, which is an ideal upper bound to control the signal variation. Research in [8,72] suggests that windows of 150–250 ms for EMG are the optimal tradeoff between the classification accuracy and the delay, while 100–250 ms windows can be used for mechanical sensors. Fast time response is also desired for continuous classification tasks in addition to high classification accuracy [29]. In [42,72], a method called overlapping analysis windows was applied to accelerate the decision update. The key aspect of this scheme lies in the careful window increment setting. The smaller the window increment step, the faster the classification result can be achieved. Furthermore, overlapping can help obtain better classification performance when processing the transition phase, such as from the sitting phase to the standing phase [42].

3.3. Feature Extraction

After the signal was preprocessed, the noise was attenuated and the window segments were obtained. However, using the entire segment as the input data is not an ideal choice due to the high computation cost and poor correlation between the input data and the target patterns, which would negatively affect the recognition performance. Therefore, it is necessary to extract some features from each window sequence. In this section, the two most commonly applied feature extraction approaches, including time-domain features, frequency domain features, and autoregressive features, are reviewed and compared.

The time-domain features of the EMG signal are based on the characteristic index of the statistical method, which regards the EMG signal as a function of time. This is an intuitive interpretation of the EMG signal. Its calculation is directly related to the magnitude of window segments, which occupies small calculation resources. In [71,73,74], Hudgins et al. proposed several time-domain features that are widely used in the following researches, including mean absolute value (MAV), mean absolute value slope, slope sign changes (SSC), waveform lengths (WL) and zero crossings (ZC). In [75], time-domain features like root mean square (RMS), waveform length (WL), number of zero-crossings(ZC), variance (VAR) and maximum(MAX) value are combined to represent the feature space.

The frequency-domain features of EMG signal are mainly to transform the time-domain signal of EMG signal into a frequency-domain signal through Fourier transform, and then analyze the spectrum characteristics or power spectrum characteristics of the signal. The advantage of this method is that it overcomes the characteristics of a time-domain signal, which is greatly affected by noise, is relatively weak, and it is easier to extract stable characteristic indexes by analyzing EMG signals in frequency-domain. Common frequency domain features are comprehensively investigated in [76]. Lots of experimental results demonstrate that mean frequency (MNF), median frequency (MDF), mean peak frequency (PKF), mean power (MNP), frequency ratio (FR), power spectrum ratio (PSR), and variance of central frequency (VCF) are not good features for EMG signal-based locomotion mode classification. Joint time-frequency domain features have been proved to be capable of effectively representing transient EMG patterns resulting from dynamic contractions [77]. The experiments comparing the time domain methods and time-frequency methods have also been conducted, and results show that the wavelet packet transform (WPT) is the best method to increase the EMG information density.

Apart from the conventional time and frequency domain feature extraction approaches, there are other ways of extracting effective features. For example, autoregressive (AR) features are usually taken into consideration in lower limb locomotion mode analysis. Huang. He et al., applied autoregression features (three-order autoregression coefficients) and time-domain features (MAV, ZC, and WL) for locomotion identification for their computation efficiency and fast time-response property.

In recent years, there are also handcrafted features that do not rely on statistical techniques. Instead, they use the prior knowledge of biomechanics to select the specific response point at the experimental curve. For example, in [43], twenty-one biomechanical features such as the peak EMG interval between two selected muscles and the duration of the EMG activation time are carefully selected. Such an extraction technique obtained an accuracy of 96.8% and is proven to outperform the traditional feature extraction technique. Although this kind of method seems lacking the general capability dealing with different scenarios, the classification performance is satisfying for the only specific task and shows less generality than the conventional features. There are also researches dedicating to improving the efficiency of the conventional feature extraction methods by designing new features. For example, in a most recent research [78], three time-domain features, including ASS,

MSR, and ASM, are specially proposed to improve the performance of EMG-PR based strategy in arm movement classification task. Experimental results showed that the new-designed features could achieve the accuracy 92.00% ± 3.11%, which is 6.49% higher than that of commonly used time-domain features [72].

3.4. Dimensionality Reduction

When the extracted feature set occupies a high-dimensional space, it may lead to the failure to correctly classify the patterns because the discriminative features are hidden in the complex feature space. Efficient data dimensionality reduction technique not only reduces the computation cost but also proves to be powerful in increasing the inter-class distance, enhancing recognition accuracy. Two commonly used dimensionality methods, i.e., feature projection and feature selection, are being discussed in this section.

3.4.1. Feature Projection

The feature projection method means projecting the original high dimension feature space into a lower-dimensional feature space. The most important characteristic of such a method is that it uses all the original features to get the reduced feature space, which will not bring any information loss. Many algorithms are proposed to accomplish this task, including approaches like principal components analysis (PCA) [77,79], Non-negative matrix factorization (NMF) [80], averaging [75], independent components analysis [81], nonlinear projection [82] and. We will mainly introduce the PCA and NMF approaches in this section.

PCA is regarded as one of the most commonly used linear dimensionality reduction techniques. It essentially takes the direction with the largest variance as the main feature and off-correlates the data in each orthogonal direction, which makes them irrelevant in different orthogonal directions. Therefore, PCA also has some limitations. For example, it can remove linear correlation very well, but there is no way for higher-order correlation. For data with high-order correlation, Kernel PCA can be considered, and the nonlinear correlation can be achieved through the Kernel function turning to linear correlation. The research in [83] illustrates the importance of having relevant embedded muscle activity features in a low-dimensional space. It is also demonstrated in this paper that the PCA technique can efficiently capture features from EMG signals in an unsupervised manner. Due to the parameterless characteristic, PCA is convenient for universal implementation but having trouble for personalized optimization itself. Besides, some other methods about the strategic combination of features have also been leveraged in recent papers, and they are improving towards the direction with higher speed and accuracy. For example, FastICA is proposed in [84] to overcome the difficulty of previously used ICA for high-density EMG signal decomposition.

Non-negative Matrix Factorization (NMF) is another dimensionality reduction technique that has also been widely studied in recent years [80,85]. D.D.Lee et al. point out in [85] that NMF reduces data dimensionality by approximately factorizing the source data into two matrices. The factorization aims to extract the latent structure from the source data to a lower-dimensional space by minimizing the predefined cost function whose value is inversely proportional to the quality of the approximation. Common cost functions include conventional least square error and Kullback-Leibler divergence. As for its application in EMG signal recognition, G.R.Naik and H.T.Nguyen first studied the identification of EMG finger movement with a nonnegative matrix factorization technique [80]. Recently, non-negative matrix factorization is also leveraged to select the subject-specific signal channel for improving the lower-limb motor imagery recognition accuracy. Apart from this, NMF is also found to be put into practice in the determination of muscle synergies while utilizing EMG signals [86].

Actually, lots of other data dimensionality reduction methods have been investigated in recent years due to the quick progress in data science, like Uncorrelated Linear Discriminant Analysis (ULDA) [87], the locality preserving projections (LPP), neighborhood preserving embedding (NPE), discriminant analysis (OFNDA) and so on. These methods provide different dimensions to increase the distinguishable information density, which will help improve recognition accuracy and processing speed.

3.4.2. Feature Selection

Feature selection means reasonably selecting a subset of features to form the new feature space, while the rest of the original features would be abandoned [88,89]. It has been proved by numerous empirical results that feature reduction can reduce data dimensionality and the computational complexity meanwhile.

A key problem in the feature selection method is that what factors should be taken into consideration to define the best EMG feature space, Zardoshti-Kermani, Mahyar, et al. [90] came up with some standards to evaluate the EMG feature space including maximum class separability, robustness, and computational complexity.

Diverse approaches have been proposed to help select more representing features from the dataset containing too much useless data. Some work derives from combining the specific physical or statistical properties. In the task like limb posture classification, deterministic approaches have been proposed to select the optimal feature-channel pairs [91]. In this work, a distance-based feature selection to determine a separability index is utilized. Besides, to measure the amount of mutual information between features and classes, a correlation-based feature selection method was applied, and both schemes were proved effective to boost the posture classification accuracy.

Evaluating relevant features is another novel approach to make feature selection proposed in recent years [92–97]. One newest work formulates the feature selection problem as the optimization problem and proposes a personal best guide binary particle swarm optimization (PBPSO) to solve the feature selection, which works by evaluating the most informative features from the original feature set [93]. In another recent paper [97], a feature selection algorithm called ReliefF is given from the heuristic angle. This algorithm first selects the random instance of one of the database classes and then searches for the k nearest neighbor instances.

3.5. Classification Algorithms

Once the features with the reduced dimensionality are determined, classification algorithms should be deployed to distinguish the different categories of the extracted feature vectors. Numerous algorithms have been proposed, and in this section, we will review some of the most frequently utilized algorithms in EMG PR, along with the comparison between them.

Traditional classification methods include linear discriminant analysis (LDA) [98], support vector machine (SVM) [99,100], k nearest neighbor (KNN) [101,102], Bayesian analysis, fuzzy logic (FL) [103] and hidden Markov models(HMM) [104] while some modern methods have also been given like artificial neural network (ANN) [105] and convolutional neural network (CNN) along with the recent progress in deep learning research. All of the above methods are effective in classifying the extracted feature space while they each have specific characteristics due to their intrinsic difference at the algorithm level. Therefore, we will briefly explain the basic principle of each type of algorithm along with the comparison of their classification performance at different rehabilitation scenarios.

LDA is one of the most simple but effective methods which has attracted great attention in recent years. Linear discriminant analysis is utilized in for lower limb prosthesis movement mode recognition and achieves competitive accuracy of 97.45% even compared with neural network classifiers while deploying a PCA reduced feature set. In another recent work [105], LDA in a One-Vs-One topology was used for non-weight bearing lower-limb movement recognition immediately after training. Nevertheless, a fatal disadvantage for the LDA model is that it is not capable of handling the linear inseparable problem even if given perfect data. At this point, ANN enjoys the ability to describe nonlinear class boundaries among different categories. MLP and Cascade are the two main structures of artificial neural networks. Identification and classification of different gait phases according to the collected EMG data is a classical linear inseparable problem. In new research focusing on this problem [104], three different MLP is deployed for the classification of two main gait phases while directly using the raw EMG data without feature extraction. The experimental comparison demonstrates that the performance of three MLP models achieves an average higher accuracy over all the popular models. On the other hand, in scenarios that some dimensionality reduction techniques like PCA have linearized the discrimination task, LDA can behave even better than the ANN methods. Besides, due to the well-defined classifier inner architecture, LDA possesses better performance stability. Nevertheless, for a given problem, it may be not easy to determine the optimal size and structure of an ANN. At last, in spite of the advantage of extracting latent features from unprocessed EMG data, the long training period is another thing that we need to take into consideration when intending to apply the ANN.

The SVM is estimated as one of the most popular approaches utilized in the movement mode classification. It is an effective data classification approach that projects the low-dimension data to the high-dimension feature space via kernel function. SVM works by finding a hyperplane to distinguish different categories in high-dimension space through the training process. It is believed that proper kernel function can help reduce the indivisible linear data to linear separable set in high-dimension space. From this, it is obvious that the most crucial problem when building the SVM model is to determine the optimal kernel function and its parameter values. Pires, Ricardo, et al. [99] uses the SVM algorithm to classify lower-extremity EMG signals during running shod/unshod with different foot strike patterns. It is also mentioned that kernel function selection, and parameter setting should depend on the specific task scenario. In [100], the SVM algorithm is utilized to develop an automatic classification system for lower limb hemiparetic patients. The empirical results demonstrate that SVM has the highest accuracy (95.2%) than KNN (89.2%) and ANN (92.3%). Particle swarm optimization (PSO) has long been a classical heuristic optimization algorithm. According to our literature review, the related works hybridizing the PSO and SVM to detect movement patterns are growing continuously in recent years. Zheng, Jiajia, et al. [106] employ PSO-optimized SVM to classify four gait phases, including initial contact, mid stance, terminal stance, and swing phase. The experimental results show that PSO-SVM exhibits the distinctive advantages on gait phase classification and improves the classification accuracy up to 32.9%–42.8% compared with the classifier based on vanilla SVM. Recently, some work compares the performance of common classifiers in different scenarios to provide references in practical application. The comparative analysis among NLR, MLP and SVM shows that for either classification performance and for the number of classification parameters, SVM attains the highest values followed by MLP, and then by NLR [107]. It should also be noted that SVM can obtain the highest classification performance while utilizing the lowest sampling rate.

Fuzzy Logic (FL) is another technique used in the classification of EMG signals, which achieves definite conclusions just from imprecise data input in a simple manner. It has been studied that FL exhibits the advantage of control techniques in biosignal processing [103]. FL can extract unrepeatable EMG features and mimic user's intent to make a decision. On the other hand, FL requires more system memory and processing time because of the use of fixed geometric-shaped membership functions in fuzzy logic limits system knowledge more in the rule base than in the membership function base [103].

The application of the hidden Markov model (HMM) is described in [108], where it is used for recognition of gait mode based on electromyographic signals. A modified Baum-Welch was used to estimate the parameter of HMM, and the Viterbi algorithm achieved the recognition of gait mode by finding the best HMM and state to assign corresponding phases to the given segments.

KNN classifier is another kind of common biosignal classifier based on traditional machine learning techniques. The authors in [101] use the K-nearest neighbor method to classify the EMG signals recorded from lower limb muscles during standing in individuals with complete spinal cord injury implanted with spinal cord epidural stimulation. Actually, most KNN-related researches in this area are correlated with other classification algorithms to find the most suitable classifiers for some certain problem. In [102], weighted KNN and SVM are both utilized as the classifier to distinguish the sEMG signals for controlling a prosthetic foot while the comparison of the accuracy of two classifies shows that weighted KNN obtains higher efficiency.

With the rapid development of deep learning technology in recent years, increasing attention has been paid to the deep feature classifier. For example, a backpropagation neural network was used in [109] to map the optimal surface EMG features to the FE angle angles. The experimental results show that the features extracted from multichannel surface EMG signals using deep belief network method proposed in this paper outperform principal components analysis (PCA), and the RMS error between the estimated joint angles and calculated ones during human walking is reduced by about 50%.

4. Multisensory Fusion

Through a series of processing, the change of movement patterns creates a stronger correlation with the EMG signal. However, due to the interactions between the lower limb movement and the external environment, as well as the low resolution of the upper limb hand fine movement in the EMG signal, the sensing strategy relying only on the EMG signal has great limitations in these two types of rehabilitation scenes. To address these issues, multisensory fusion techniques are proposed and proved to be capable of boosting pattern recognition accuracy by collecting more efficient data from different dimensions.

Generally, the EMG-centered fusion strategy can be divided into two types, i.e., fusion with kinematic sensors and fusion with kinetic sensors. The former strategy is mainly to supplement the three-dimensional motion information like human body acceleration, angular velocity, angle, etc. While the latter one is mainly to supplement the force information during the movement process. In this section, we will explain the detailed procedures of these two fusion strategies and demonstrate some successful applications using these techniques in different rehabilitation scenarios. Figure 7 demonstrates a block diagram of the above EMG-centered fusion strategy.

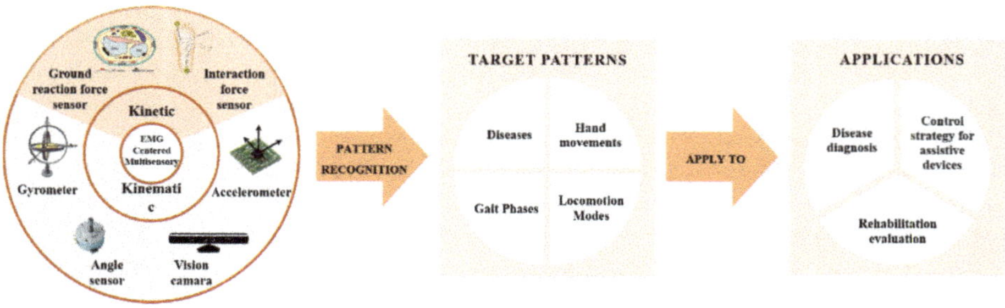

Figure 7. EMG-centered multisensory strategies and sensing blocks.

4.1. Fusion with Kinematic Sensors

Kinematic sensors are widely used in obtaining the locomotion properties of targets. In recent years, kinematic sensors have also attracted enough attention in human behavioral study combined with biosignal sensors (e.g., EMG). Kinematic information of human lower extremity describes the movements of the joints and limbs. Generally, the kinematic sensors can be divided into two main categories, inertial sensors and motion capture systems. Inertial sensors, including accelerometer, gyroscope, and magnetometer, are often used to collect kinematic data like joint angles, joint angular velocity, body orientation, and limb acceleration, etc. Motion capture systems became popular only in recent years, their biggest advantage being that they can collect locomotion information in a contactless manner.

Kinematic sensors' advantage of intuitively reflecting the 3D motion information of the limb helps compensate for the defect of EMG sensors, which only describes the whole physical process from the human nerve activity angle. Combing bioelectrical signal and motion information like acceleration has also been investigated to improve locomotion classification accuracy. Researchers in [110] especially investigate the improvement in classification error while utilizing EMG and kinematic data together rather than pure kinematic data. The empirical results demonstrate that the higher recognition rates can be obtained with combined data compared with vanilla kinematic information from the hip joint of non-dominant/impaired limb and an accelerometer.

The most successful use of the kinematics-EMG fusion strategy lies in the control of powered lower-limb prostheses. Such devices help people get rid of the disability and restore normal walking ability. Nevertheless, the adaptability of the prosthesis to other parts of the human body is still a challenging subject in which the kernel issue can be regarded as inferring human locomotion intent. Due to the restriction of the traditional recognition approach only relying on kinematic data, some recent works are making the steps to fuse EMG and mechanical information to improve the performance. The locomotion recognition system used for lower limb prosthesis control in [35] fuse the multichannel EMG signals and accelerometer measurements in feature level. The physical test shows that the empirical accuracy is over 95%, which is significantly better than traditional schemes, only applying surface Electromyography for identifying locomotion modes [29,35]. In [111], A combination of kinematic and electromyographic (EMG) signals recorded from a person's proximal humerus was used to evaluate a novel transhumeral prosthesis controller. Most especially, they trained a time-delayed artificial neural network to predict elbow flexion/extension and forearm pronation/supination from six proximal EMG signals, and humeral angular velocity and linear acceleration. Young, A. J. [15] studied the contribution of EMG data in combination with a diverse array of mechanical sensors to locomotion mode intent recognition in transfemoral amputees using powered prostheses. And this can indeed significantly reduce intent recognition errors both for transitions between locomotion modes and steady-state locomotion.

In addition, EMG centered kinematic fusion technique is also seen in applications of the hand gesture recognition. This is because the dexterous movement of the finger movements varies more significantly in the position information than the EMG information. Therefore, it is an ideal choice to combine them to obtain intention information and position information at the same time. For example, inertial measurement unit(IMU) and myoelectric units are utilized to implement the hand gesture-based control of an omnidirectional wheelchair in [112]. The system component, sensor placement, and the characteristic collected signals are demonstrated in Figure 8. In this system, the classification, which involves recognizing the activity pattern based on the periodic shape of triaxial wrist tilt angle and EMG-RMS from the two selected muscles, helps the accuracy improve to 94%.

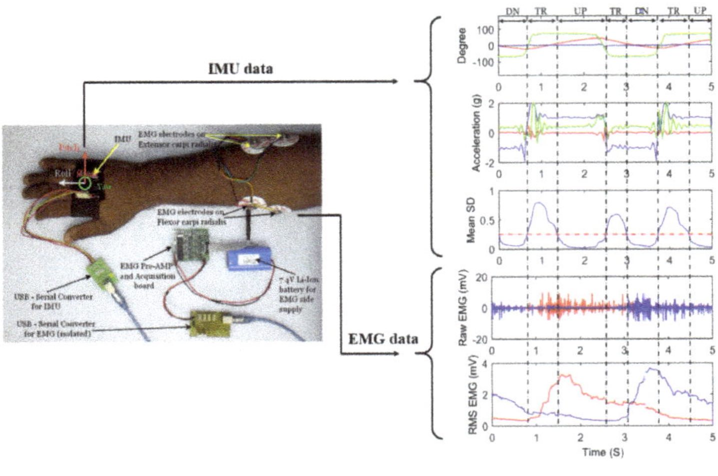

Figure 8. The fusion strategy of combining kinematics data (IMU) and EMG data, adopted from [112].

Utilizing vision technique is another effective and direct way to detect the kinematic information of human. Recently, with the vigorous development of computer vision, some researchers are trying combining the vision and EMG information for limb motion classification task. In a newly published work [113], researches implemented a framework that allows the integration of multi-sensors, EMG and visual information, to perform sensor fusion and to improve the accuracy of hand gesture recognition tasks. For embedded applications, even-based cameras were utilized to run on the limited computational resources of mobile phones. The online results of hand gesture recognition using fusion approach reached 85%, which is 13% and 11% higher than utilizing EMG and vision individually.

4.2. Fusion with Kinetic Sensors

Kinetic sensors are a kind of instrument used to measure forces and moments that are directly connected with the movement of body segments like lower limbs. Various types of force transducers, including piezoelectric [114,115], strain gauged [116,117], and capacitive transducers [118–120], have been widely used in the design of force-sensitive sensors. Joint use of kinetic sensors and electromyography sensors is believed to be able to extract human muscle reflex related information, which will help the recognition of specific body behaviors. Generally, two types of kinetic information, interaction force, and ground reaction force can both be integrated with EMG information. The following part will introduce the fusion examples with interaction force and ground reaction force, respectively.

Interaction force sensors are typically embedded in a fixed structure and then used to detect the extension or flexion of limb, which are important for interpreting the intention of the human movement. The utilization of interaction force and EMG signals are also seen in the application of the upper limb hand gesture recognition in recent years. The research in [27] showed that combining EMG and pressure data sensed only at the wrist could support the accurate classification of hand gestures. Especially, the EMG is suited to sensing finger movements, the pressure is suited to sensing wrist and forearm rotations, and their combination is significantly more accurate for a range of gestures than either technique alone. Figure 9 demonstrates the implementation of this multisensory approach in hand gesture recognition.

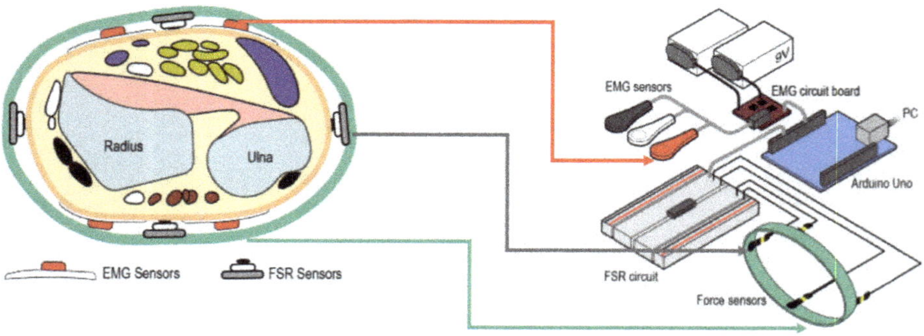

Figure 9. EMG-FSR fusion strategy and hardware diagram used to classify different hand gestures, adopted from [27].

According to our literature review, ground reaction force (GRF) is also a commonly used kinetic information, especially in lower limb rehabilitation scenarios. A typical combination pipeline focus on lower limb prosthesis is shown in Figure 10. Such integration strategy is mainly applied in the lower limb exoskeleton or prosthesis because the lower limb is mostly responsible for interaction with the ground, and many locomotion modes are closely related to the terrain changes, which is directly reflected in the GRF information. Liu, Ming. et al. [35] applied the EMG measurement results and mechanical ground reaction forces/moments recorded using a six-degree-of-freedom load cell to build an adaptive classification strategy, which further enhanced the reliability of neutrally-controlled prosthetic legs. The researchers in [60] presented a smart sensing system utilizing flexible electromyography sensors and ground reaction force sensors for locomotion mode recognition. Here, EMG and GRF information were collected from ten healthy subjects in five common locomotion modes in daily life. Rehabilitation robots is an attractive research zone which mainly helps the patients like stroke regain the locomotion ability. In [121], ground reaction forces and lower extremity electromyography are both utilized to compare inclined treadmill walking and turning conditions with its emulation on a developed balance assessment robot in order to investigate the feasibility of the existing approaches in rehabilitation.

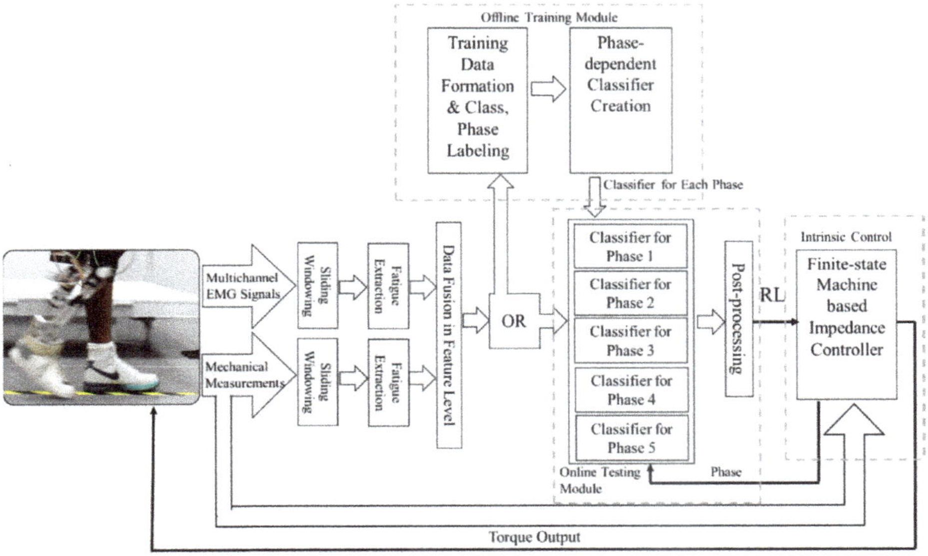

Figure 10. EMG-GRF fusion strategy and the fusion pipeline to recognize different locomotion modes adapted from [27].

4.3. Fusion with Both Kinematics and Kinetic Sensors

In addition to the information fusion work introduced above, we also notice some recent work combing EMG with both kinematic and kinetic data. With the improvement of processor performance and the development of machine learning algorithms to handle high dimensional data, this triple-type sensor information fusion, which used to be regarded as a difficult approach, has been implemented and outperforms existing approaches in some motion classification tasks.

Rehabilitation exoskeleton is an important application of kinematic-kinetic fusion strategy. Because this kind of equipment needs various dimensions of information to maintain the highest stability and safety, it can achieve the goal of using EMG as the subjective control input and kinematic and dynamic data as the feedback. The work in [122] used human-robot interaction force, ground reaction force, and EMG sensors to distinguish the walking environment and gait period to provide crucial information for exoskeleton control. The results obtained using an individual mechanical sensor together with sEMG showed improvement compared to the case of only using force sensors, reaching the highest accuracy of 97.8% at gait phase recognition scenarios. The sensing blocks, as well as the location of the sensors, have been demonstrated in Figure 11.

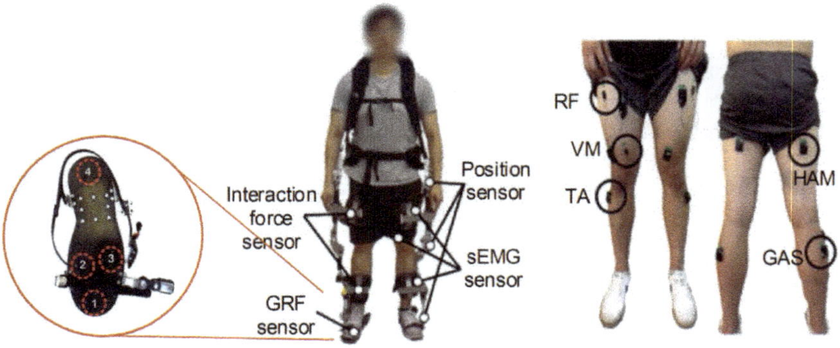

Figure 11. Multisensory fusion strategy using EMG, kinematic sensors, and kinetic sensors together for the application of rehabilitation exoskeleton, adopted from [122].

Gait pattern analysis has also been widely used to assist in the diagnosis of muscle diseases. By multi-sensor data fusion, gait analysis gives objective and quantitative information about the gait pattern and the deviation due to the muscular situation of these patients, which are of great significance to the clinical research. Cui, Chengkun, et al. [121] conducted the simultaneous recognition and assessment of post-stroke hemiparetic gait by fusing kinematic, kinetic, and electrophysiological data. In [123], researchers investigated the gait pattern of 10 patients with myotonic dystrophy (Steinert disease) compared to 20 healthy controls through manual muscle test and gait analysis in terms of kinematic, kinetic and EMG data.

Table 2 summarizes the fusion strategy, targeted recognition classes, extracted features, proposed classifiers, and the final accuracy results of the multisensory application in different scenarios of rehabilitation.

Table 2. Summary of the pattern recognition techniques applied in relevant studies.

No.	Applied Sensors	Classes	Feature	Classifier	Accuracy
[35]	EMG + GRF	Five common locomotion modes (W, RA, RD, SA, SD) and eight task transitions: W->SA, W->SD, W->RA, W->RD, SA->W, SD->W, RA->W and RD->W	EMG data feature: MAV, SL, SSC and ZC, mechanical signals: maximum, minimum, mean value and standard deviation	Entropy-based adaptation (EBA), Learning form testing data (LIFT) and Transductive Support Vector Machine (TSVM)	EBA: 95%, LIFT: 95% and TSVM: 96.25%, vanilla SVM: 87.5%
[42]	EMG + GRF	Locomotion modes: LW, SO, SA, SD, RA and RD and related transitions: W->sA, W->RA, W->O, SD->W, RD->W, SA/RA->W, W->SD/RD	EMG time-domain feature: MAV, SSC, WL, ZC, Mechanical signal features: maximum, minimum, mean value of each direction of force and moment	SVM	99% or higher accuracy in the stance phase and 95% accuracy in the swing phase

Table 2. *Cont.*

No.	Applied Sensors	Classes	Feature	Classifier	Accuracy
[122]	Position sensors, GRF, interaction force EMG	Five walking environments: LW, RA, RD, SA, and SD Seven gait periods: LS, MST, TST, PS, IS, MS and TS	GRF feature: four positions in the foot for four time periods, position feature: three joint angles for four time periods. Interaction force feature: two points in the link for four time periods, sEMG feature: MAV, ZC, SSC and WL	BLDA	96.1%(environment classification accuracy) 97.8%(gait period classification period)
[27]	EMG sensor, pressure force sensor	Finger gestures, wrist gestures, and other gestures	Root mean square (RMS), standard deviation (SD) and peak amplitude	SVM	95.8%
[112]	IMU, EMG sensor	Six hand gestures (forward, clockwise, left, backward, anticlockwise, right)	Nine IMU features extracted from wrist Euler angle and six EMG features extracted from EMG RMS signal	DSVM	Real-time recognition accuracy 90.5%
[124]	EMG signal acquisition system, data glove	Thumb flexion, finger flexion, thumb opposition, middle/ring/little finger flexion, long fingers flexion, tradigital grasp, lateral grip/key grip	MAV (mean of absolute value)	Locally weighted learning	79% for amputee and 89% for non-disabled participants

Classes: LW = level-ground walking; RA = ramp ascent; RD = ramp descent; SA = stairs ascent; SD = stairs descent; SO = stepping over an obstacle; IT = ipsilateral turning; CT = contralateral turning; SS = standing still; LP = loading response; MST = midstance; TST = terminal stance; PS = pre-swing; IS = initial swing; MS = mid-swing; TS = terminal swing. Feature: MAV = mean absolute value; ZC = zero crossing; SSC = slope sign changes; WL = waveform length; SL = signal length.

5. Challenges and Future Development

Currently, fruitful results have been achieved by EMG-centered multisensory pattern recognition methods in the rehabilitation field. However, there are still limitations in terms of accuracy, robustness, data volume, and flexibility when current approaches are applied to practical applications. In this section, we will explain the main issues and potential solutions and research directions of multisensory EMG PR.

5.1. Low Data Quality

Low data quality is a pressing problem in EMG detection, especially under a multisensory fusion scenario. Because the data quality of each dimension can significantly affect the final recognition accuracy. Since bioelectrical signals are generally more susceptible to noise interference, in EMG centered multisensory fusion, the data quality of EMG signals is the most important influential factor.

In Section 3, we explain the preprocessing of the EMG signal, whose purpose is mainly to reduce the noise in the raw EMG signal. Although by the preprocessing technique, the deterministic noise, e.g., charger noise, can be filtered and stochastic noise such as white noise can be compressed, interference comes from the source of the bioelectric signal that is challenging to be removed by data processing technique. First, the EMG crosstalk effect is an inevitable interference in EMG detection [22].

It means that the amplitude of the EMG signal is not only contributed by the targeted muscle but also influenced by the adjacent and surrounding muscles, hindering precise detection of the muscle activity. Second, electrode shift also results in severe EMG signal misregistration, since the traditional electrochemical gel electrode suffers difficulty in firmly attaching to the skin; a shift of merely 1 cm would reduce the classification accuracy by 5% to 20% [125]. Third, even when the patient performs the same locomotion under the same environment, the applied muscle force may slightly change due to the impossibility in perfectly controlling the body, resulting in the variations of the EMG signals of the selected muscles, and reducing the data quality [19].

Advanced sensing and detecting means could be possible solutions to the problem of low-quality data. With the integration and miniaturization of sensory circuits, high-density EMG (HD-EMG) has become a promising sensing approach. This type of sensing technique increases the coverage and density of the electrodes. Hundreds of electrodes are integrated with a small size overlying a restricted area of the skin [126,127]. The most significant benefit of HD-EMG is that the possibility of subtracting information at the motor unit level greatly improved compared with conventional EMG, minimizing the effect of EMG crosstalk at this precision level. The multiple electrodes are arranged in the form of a two-dimensional array, for example, in the shape of 8×24 [127] or 8×16 [126], which will produce an "EMG image". The EMG image presents a comprehensive view of EMG signals both in the time domain and spatial domain, providing far richer information than the conventional EMG electrode placement method of implementing one or two electrodes at a single muscle. For example, the HD-EMG arrays attached to the elbow area were used for the precise hand gesture recognition, reaching the accuracy as high as 99.0% in classifying twenty-seven gestures [126]. A typical procedure of utilizing the "EMG image" generated by the HD-EMG to conduct pattern recognition is demonstrated in Figure 12. Even though abundant information can be extracted with the help of HD-EMG, it also increased the burden of the acquisition and processing hardware. Because the sampling rate of EMG signals requires at least 1000 Hz at each channel, an 128 channel HD-EMG system will produce over 128,000 observations in one second, which would make the data scale enormous and lead to challenges in real-time processing.

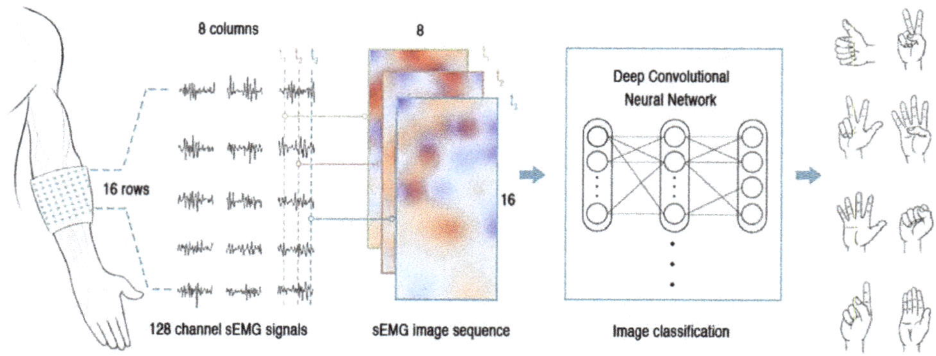

Figure 12. The procedure of applying High-Density EMG (HD-EMG) in discriminating hand gestures, adopted from [126].

5.2. *Inadequate and Undisclosed Data*

High-quality data acquisition is the premise of high-precision multisensory pattern recognition. Meanwhile, for multi-sensor fusion, due to the large variability in the patient's muscle and physical condition and the difficulty in obtaining high-quality multisensory data, adequate and open-source data is of equal importance. This section will explain the necessity of establishing a big and open sensor data library in detail, and review some currently developing databases.

Currently, most of the current EMG researches were recruiting a relatively small volume of the volunteers to collect EMG signals, such as 5–20 healthy individuals or no more than five patients. The relatively small size is owing to the limited access to experimental resources due to the insufficient funding budget. Therefore, recruiting tens of participants is a balanced choice for verifying the proposed algorithms for most research groups. Nevertheless, there are a few concerns about this choice.

Firstly, the variation of the EMG signal is usually significant in different individuals. A strong correlation between EMG signals and the age, height, weight, exercise habits, and health status of individuals are found in relevant studies [128]. The sample size of tens of experimenters is often difficult to cover all these differences between various populations. Besides, researchers usually adopt their experimental protocols and different sensing equipment, and it is not easy to investigate the generality of the method proposed by one work across these influential factors.

Furthermore, the requirement of sensing equipment for robust EMG acquisition is quite high, and the collection system is typically expensive. It may potentially hinder some labs or researchers who do not have the experimental conditions from investigating EMG signals on the algorithm level.

An important approach for solving the above problem is to promote the concept of "big data-based EMG," which means that the verification method of EMG pattern recognition should shift from the verification in small sample experimental data to the performance of the proposed algorithm based on big data in different populations. Well-established datasets not only allows researchers to examine the quality and correctness of their acquired data more conveniently but testing the generality of their proposed methods in different datasets. In addition, larger datasets enable the application of some advanced algorithms such as deep learning in digging the underlying information in EMG signals [129].

In the past, the experiment data of EMG were mostly stored within the laboratories of individual researchers rather than publishing online, which made it hard to gather multiple datasets. In recent years, with the occurrence of "data paper," researchers are encouraged to publish their experimental data as an individual publication [130]. Such scientific publications promote the transparency of the raw data, and thus more EMG datasets are being available. In addition, there are also studies dedicated to establishing a standard dataset, trying to provide a standard experimental procedure and cover as many different muscles, movements, and population as possible. A few influential datasets such as Surface Electromyography for the Non-Invasive Assessment of Muscles (SENIAM), The Ninapro project [131] have emerged in recent years. However, there is still a long way to go in reaching the consensus of a widely accepted protocol and establishing a thorough dataset.

5.3. Discrete Interpretation of Continuous Movements

The development of the data quality and quantity offers the possibility to develop more effective processing techniques, especially in improving the simultaneity and continuity of the EMG-PR-based control strategy, which are two key drawbacks at the state of the art technique. In this section, we will explain how simultaneity and continuity problem is affecting the EMG PR performance, and provide possible solutions and research directions in the light of the multisensory technique.

As mentioned earlier, the EMG PR technique is widely used in the real-time Human-Machine Interface system (HMI). However, the broad use of clinical viable EMG based robots are still not reported yet. The potential problem is that a complete pipeline, including sensing, signal transmission, and processing, has to be done in order to recognize the current pattern. There is a time delay in this procedure, and it is acceptable in some scenarios, such as sign language recognition, when only static patterns like hand gestures are to be recognized. However, the patterns relevant to lower limbs are mostly dynamic and require real-time and instant recognition, and therefore the time delay may greatly influence the application of conventional PR technique in lower limb HMI devices.

There are several attempts to resolve the problem of discrete interpretation. One of them is to design the transition logic between two consecutive patterns skillfully, and some pioneers studies are trying to design the transition strategy. In [42], overlapped windows are utilized to overcome the time delay problem, and the comparison between the overlapped windows and non-overlapping windows are depicted in Figure 13. The increment of two consecutive and the overlapped window is only 12 ms, generating approximate continuous decision flow. The majority vote method is used to prevent misjudging problem. In addition, because the EMG signal is generated before the actual motion, there are also studies using the currently collected signals to predict the patterns in the next moment so that the robotic actuators could react in advance to adapt to the changing patterns [132,133]. However, the prediction method is only applicable in the periodical movement like gait, and still have difficulties in predicting more complicated motions. So far, there is still no consensus on the best transition method among EMG patterns, and the robust transition is a research interest in the EMG PR field.

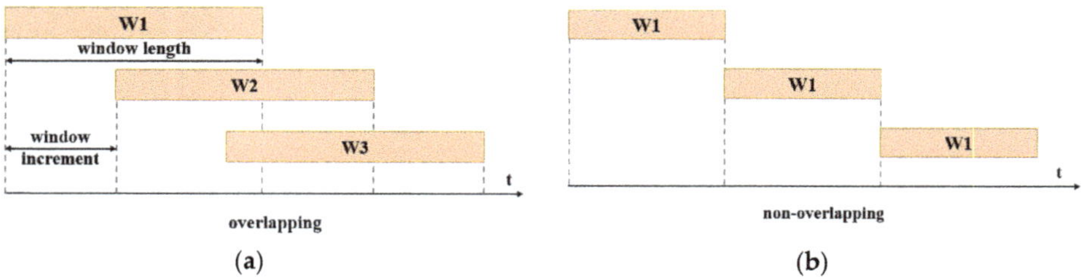

Figure 13. A conceptual depiction of: (**a**) the overlapping windowing method. (**b**) the non-overlapping windowing method.

With the development of multisensory fusion and the increasingly thorough study of human body motion, the method of combining the biomechanical model with pattern recognition is also attracting increasing attention. Some open-source software, like OpenSim, is developed to provide present biomechanical models. With the help of the multisensory fusion, both the EMG signals and other mechanical signals collected from the patient's body can be used to calibrate and specialize the raw biomechanical models. After that, with proper mapping strategies, these models can be driven by EMG signals and, therefore, intuitively simulate the movement of human extremities, providing crucial information for continuous control of HMI systems. For example, a model-based control strategy is proposed in [134], using the EMG signal as the input for the carefully tuned and calibrated hand-finger biomechanic model, and managed to continuously control a prosthesis hand to finish the task of gripping in the response time of 16.2 ms per loop. The control schemes and multisensory strategy is demonstrated in Figure 14. Nevertheless, the model-based method requires tedious procedures to calibrate and tune the model, and it is proved difficult to separate the EMG signals into distinct functional movements for the subject with small arms. At present, it is still in the laboratory research stage.

5.4. Future Analysis

In the above three sections, we explain the challenges of multisensory EMG PR and the emerging technologies potentially capable to solve these challenges. We believe that in the foreseeable future, multisensory EMG PR has high possibility to develop in these three directions for the following reasons. Firstly, the application of the novel EMG sensing technique, e.g., High-Density EMG, will greatly improve the EMG signal's quality. Besides, the spatial resolution of the EMG signals will also be enhanced from the level of distinguishing different muscle tissues to the level of distinguishing single cells, making it possible for EMG signals to apply in detection of very fine movements. Secondly, more open source databases with high data quality, complete muscle measurement points and rich application scenarios will be built, which will further reduce the obstacles of self-designed experiments and attract more researchers to join the field of EMG signal analysis. Thirdly, with the reduction of EMG research barriers, the research of EMG analysis algorithm will continue to increase, and more advanced algorithms to solve the traditional PR problem, such as the overlapping window and biomechanical model mentioned in this paper, will be proposed and verified.

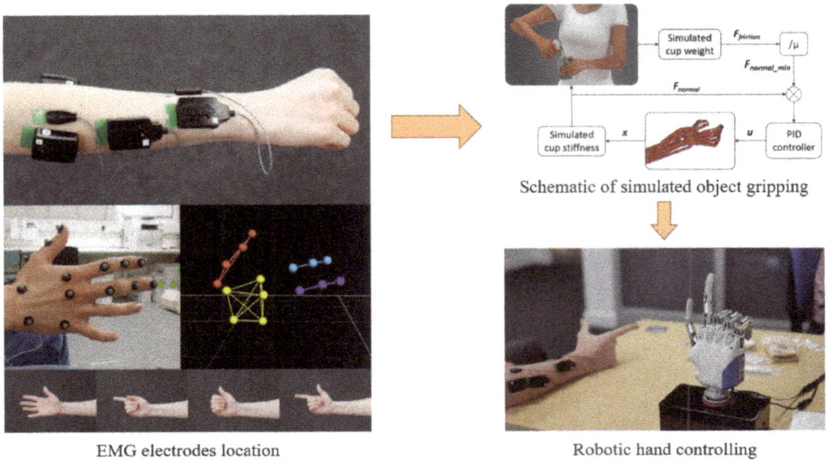

Figure 14. The schematic of utilizing biomechanical based and EMG centered multisensory technique in controlling a robotic hand, adopted from [134].

6. Conclusions

Employing electromyography (EMG) centered multisensory integration pattern recognition (PR) techniques demonstrates strong potential in precisely interpreting patients' intentions and locomotion modes, which are of significance for analyzing the progress of rehabilitation status. However, commercialized products widely applied in hospitals or home-based internet of health things (IoHT) have not been reported yet. This article aims to reveal the reason behind this phenomenon by thoroughly reviewing the state-of-the-art research outcomes and providing a comprehensive analysis of limitations. The content provided here not only allows readers to have a full picture view of this area but also shows potential directions in further enhancing the development of applying EMG centered multisensory fusion PR techniques in rehabilitation associated applications.

Author Contributions: Conceptualization, C.F., and S.G.; methodology, C.F. and B.H.; validation, C.F., B.H., and S.G.; investigation, B.H.; resources, S.G., J.C.; data curation, B.H.; writing—original draft preparation, B.H.; writing—review and editing, C.F.; visualization, Y.W.; supervision, S.G.; project administration, C.F.; funding acquisition, S.G. All authors have read and agreed to the published version of the manuscript.

Funding: This work was supported in part by the National Natural Science Foundation under Grant 61803017 and Grant 61827802, and in part by the Beihang University under Grant KG12090401 and Grant No. ZG216S19C8.

Conflicts of Interest: The authors declare no conflict of interest.

References

1. Bisio, I.; Delfino, A.; Lavagetto, F.; Sciarrone, A. Enabling IoT for in-home rehabilitation: Accelerometer signals classification methods for activity and movement recognition. *IEEE Internet Things J.* **2016**, *4*, 135–146. [CrossRef]
2. Moore, Z.; Sifferman, C.; Tullis, S.; Ma, M.; Proffitt, R.; Skubic, M. Depth Sensor-Based In-Home Daily Activity Recognition and Assessment System for Stroke Rehabilitation. In Proceedings of the 2019 IEEE International Conference on Bioinformatics and Biomedicine (BIBM), San Diego, CA, USA, 18–21 November 2019; pp. 1051–1056.

3. Nazmi, N.; Rahman, M.A.A.; Yamamoto, S.I.; Ahmad, S.A. Walking gait event detection based on electromyography signals using artificial neural network. *Biomed. Signal. Process. Control.* **2019**, *47*, 334–343. [CrossRef]
4. Joshi, D.; Hahn, M.E. Terrain and Direction Classification of Locomotion Transitions Using Neuromuscular and Mechanical Input. *Ann. Biomed. Eng.* **2016**, *44*, 1275–1284. [CrossRef] [PubMed]
5. Tao, W.; Liu, T.; Zheng, R.; Feng, H. Gait Analysis Using Wearable Sensors. *Sensors* **2012**, *12*, 2255–2283. [CrossRef] [PubMed]
6. Martinez-Hernandez, U.; Mahmood, I.; Dehghani-Sanij, A.A. Simultaneous Bayesian recognition of locomotion and gait phases with wearable sensors. *IEEE Sens. J.* **2017**, *18*, 1282–1290. [CrossRef]
7. Young, A.J.; Ferris, D.P. State of the Art and Future Directions for Lower Limb Robotic Exoskeletons. *IEEE Trans. Neural Syst. Rehabil. Eng.* **2017**, *25*, 171–182. [CrossRef]
8. Long, Y.; Du, Z.-J.; Wang, W.-D.; Zhao, G.-Y.; Xu, G.-Q.; He, L.; Mao, X.-W.; Dong, W. PSO-SVM-based online locomotion mode identification for rehabilitation robotic exoskeletons. *Sensors* **2016**, *16*, 1408. [CrossRef]
9. Kang, I.; Kunapuli, P.; Hsu, H.; Young, A.J. Electromyography (EMG) Signal Contributions in Speed and Slope Estimation Using Robotic Exoskeletons. In Proceedings of the 2019 IEEE 16th International Conference on Rehabilitation Robotics (ICORR), Toronto, ON, Canada, 24–28 June 2019; pp. 548–553.
10. Ziegier, J.; Gattringer, H.; Mueller, A. Classification of gait phases based on bilateral emg data using support vector machines. In Proceedings of the 2018 7th IEEE International Conference on Biomedical Robotics and Biomechatronics (Biorob), Enschede, The Netherlands, 26–29 August 2018; pp. 978–983.
11. Gui, K.; Liu, H.; Zhang, D. A Practical and Adaptive Method to Achieve EMG-Based Torque Estimation for a Robotic Exoskeleton. *IEEE Asme Trans. Mechatron.* **2019**, *24*, 483–494. [CrossRef]
12. Setiawan, J.D.; Ariyanto, M.; Munadi, M.; Mutoha, M.; Glowacz, A.; Caesarendra, W. Grasp Posture Control of Wearable Extra Robotic Fingers with Flex Sensors Based on Neural Network. *Electronics* **2020**, *9*, 905. [CrossRef]
13. Kańtoch, E. Recognition of sedentary behavior by machine learning analysis of wearable sensors during activities of daily living for telemedical assessment of cardiovascular risk. *Sensors* **2018**, *18*, 3219. [CrossRef]
14. Jin, D.; Yang, J.; Zhang, R.; Wang, R.; Zhang, J. Terrain identification for prosthetic knees based on electromyographic signal features. *Tsinghua Sci. Technol.* **2006**, *11*, 74–79. [CrossRef]
15. Young, A.; Kuiken, T.; Hargrove, L. Analysis of using EMG and mechanical sensors to enhance intent recognition in powered lower limb prostheses. *J. Neural Eng.* **2014**, *11*, 056021. [CrossRef]
16. Chen, X.; Zeng, Y.; Yin, Y.H. Improving the Transparency of an Exoskeleton Knee Joint Based on the Understanding of Motor Intent Using Energy Kernel Method of EMG. *IEEE Trans. Neural Syst. Rehabil. Eng.* **2017**, *25*, 577–588. [CrossRef] [PubMed]
17. Kilicarslan, A.; Prasad, S.; Grossman, R.G.; Contreras-Vidal, J.L. High accuracy decoding of user intentions using EEG to control a lower-body exoskeleton. Proceedings of THE 2013 35th annual international conference of the IEEE Engineering in Medicine and Biology Society (EMBC), Osaka, Japan, 3–7 July 2013; pp. 5606–5609.
18. Scheme, E.; Englehart, K. Electromyogram pattern recognition for control of powered upper-limb prostheses: State of the art and challenges for clinical use. *J. Rehabil. Res. Dev.* **2011**, *48*, 643–660. [CrossRef] [PubMed]
19. Samuel, O.W.; Asogbon, M.G.; Geng, Y.; Al-Timemy, A.H.; Pirbhulal, S.; Ji, N.; Chen, S.; Fang, P.; Li, G. Intelligent EMG pattern recognition control method for upper-limb multifunctional prostheses: Advances, current challenges, and future prospects. *IEEE Access* **2019**, *7*, 10150–10165. [CrossRef]
20. Kawamoto, H.; Taal, S.; Niniss, H.; Hayashi, T.; Kamibayashi, K.; Eguchi, K.; Sankai, Y. Voluntary Motion Support Control of Robot Suit HAL Triggered by Bioelectrical Signal for Hemiplegia. In Proceedings of the 2010 Annual International Conference of the IEEE Engineering in Medicine and Biology Society, Buenos Aires, Argentina, 31 August–4 September 2010; IEEE: New York, NY, USA, 2010; pp. 462–466. [CrossRef]
21. Kawamoto, H.; Lee, S.; Kanbe, S.; Sankai, Y. Power assist method for HAL-3 using EMG-based feedback controller. In Proceedings of the 2003 IEEE International Conference on Systems, Man and Cybernetics, Washington, DC, USA, 5–8 October 2003; pp. 1648–1653.
22. Solomonow, M.; Baratta, R.; Bernardi, M.; Zhou, B.; Lu, Y.; Zhu, M.; Acierno, S. Surface and wire EMG crosstalk in neighbouring muscles. *J. Electromyogr. Kinesiol.* **1994**, *4*, 131–142. [CrossRef]

23. Al-Timemy, A.H.; Khushaba, R.N.; Bugmann, G.; Escudero, J. Improving the performance against force variation of EMG controlled multifunctional upper-limb prostheses for transradial amputees. *IEEE Trans Neural Syst. Rehabil. Eng.* **2015**, *24*, 650–661. [CrossRef]
24. Sadikoglu, F.; Kavalcioglu, C.; Dagman, B. Electromyogram (EMG) signal detection, classification of EMG signals and diagnosis of neuropathy muscle disease. *Procedia Comput. Sci.* **2017**, *120*, 422–429. [CrossRef]
25. Scheme, E.; Fougner, A.; Stavdahl, Ø.; Chan, A.D.; Englehart, K. Examining the adverse effects of limb position on pattern recognition based myoelectric control. In Proceedings of the 2010 Annual International Conference of the IEEE Engineering in Medicine and Biology, Buenos Aires, Argentina, 31 August–4 September 2010; pp. 6337–6340.
26. Arozi, M.; Caesarendra, W.; Ariyanto, M.; Munadi, M.; Setiawan, J.D.; Glowacz, A. Pattern Recognition of Single-Channel sEMG Signal Using PCA and ANN Method to Classify Nine Hand Movements. *Symmetry* **2020**, *12*, 541. [CrossRef]
27. McIntosh, J.; McNeill, C.; Fraser, M.; Kerber, F.; Löchtefeld, M.; Krüger, A. EMPress: Practical hand gesture classification with wrist-mounted EMG and pressure sensing. In Proceedings of the 2016 CHI Conference on Human Factors in Computing Systems, San Jose, CA, USA, 7–12 May 2016; pp. 2332–2342.
28. Shultz, A.H.; Goldfarb, M. A Unified Controller for Walking on Even and Uneven Terrain With a Powered Ankle Prosthesis. *IEEE Trans. Neural Syst. Rehabil. Eng.* **2018**, *26*, 788–797. [CrossRef]
29. Huang, H.; Kuiken, T.A.; Lipschutz, R.D. A strategy for identifying locomotion modes using surface electromyography. *IEEE Trans. Biomed. Eng.* **2008**, *56*, 65–73. [CrossRef] [PubMed]
30. Gavin, J.P.; Immins, T.; Wainwright, T. Stair negotiation as a rehabilitation intervention for enhancing recovery following total hip and knee replacement surgery. *Int. J. Orthop. Trauma Nurs.* **2017**, *25*, 3–10. [CrossRef] [PubMed]
31. Rissanen, S.M.; Kankaanpää, M.; Meigal, A.; Tarvainen, M.P.; Nuutinen, J.; Tarkka, I.M.; Airaksinen, O.; Karjalainen, P.A. Surface EMG and acceleration signals in Parkinson's disease: Feature extraction and cluster analysis. *Med. Biol. Eng. Comput.* **2008**, *46*, 849–858. [CrossRef]
32. Jenkins, M.; Almeida, Q.; Spaulding, S.; Van Oostveen, R.; Holmes, J.; Johnson, A.M.; Perry, S. Plantar cutaneous sensory stimulation improves single-limb support time, and EMG activation patterns among individuals with Parkinson's disease. *Parkinsonism Relat. Disord.* **2009**, *15*, 697–702. [CrossRef] [PubMed]
33. Fang, Y.; Liu, H.; Li, G.; Zhu, X. A multichannel surface EMG system for hand motion recognition. *Int. J. Hum. Robot.* **2015**, *12*, 1550011. [CrossRef]
34. Yoshikawa, M.; Mikawa, M.; Tanaka, K. Real-time hand motion estimation using EMG signals with support vector machines. In Proceedings of the 2006 SICE-ICASE International Joint Conference, Busan, Korea, 18–21 October 2006; pp. 593–598.
35. Ming, L.; Fan, Z.; Helen, H.H. An Adaptive Classification Strategy for Reliable Locomotion Mode Recognition. *Sensors* **2017**, *17*, 2020. [CrossRef] [PubMed]
36. Chen, B.J.; Wang, Q.N.; Wang, L. Adaptive Slope Walking With a Robotic Transtibial Prosthesis Based on Volitional EMG Control. *IEEE Asme Trans. Mechatron.* **2015**, *20*, 2146–2157. [CrossRef]
37. Bhattacharyya, A.; Roy, S.; Sarkar, A.D. Active Exoskeleton Hand Exerciser Using Multisensory Feed Back. *Int. J. Ind. Eng. Manag.* **2017**, *2*, 33–36. [CrossRef]
38. Nazmi, N.; Rahman, A.; Azizi, M.; Yamamoto, S.-I.; Ahmad, S.A.; Zamzuri, H.; Mazlan, S.A. A review of classification techniques of EMG signals during isotonic and isometric contractions. *Sensors* **2016**, *16*, 1304. [CrossRef]
39. Farina, D.; Merletti, R.; Stegeman, D. Biophysics of the generation of EMG signals. *Electromyogr. Physiol. Eng. Noninvasive Appl.* **2004**, *4*, 81–105.
40. Ibrahim, A.; Gannapathy, V.; Chong, L.; Isa, I. Analysis of electromyography (EMG) signal for human arm muscle: A review. In *Advanced Computer and Communication Engineering Technology*; Springer: Berlin/Heidelberg, Germany, 2016; pp. 567–575.
41. David, A.W. *The Biomechanics and Motor Control of Human Gait*; University of Waterloo Press: Waterloo, ON, Canada, 1988.
42. Huang, H.; Zhang, F.; Hargrove, L.J.; Dou, Z.; Rogers, D.R.; Englehart, K.B. Continuous locomotion-mode identification for prosthetic legs based on neuromuscular–mechanical fusion. *IEEE Trans. Biomed. Eng.* **2011**, *58*, 2867–2875. [CrossRef] [PubMed]

43. Gao, S.; Wang, Y.; Fang, C.; Xu, L. A Smart Terrain Identification Technique Based on Electromyography, Ground Reaction Force, and Machine Learning for Lower Limb Rehabilitation. *Appl. Sci.* **2020**, *10*, 2638. [CrossRef]
44. Lay, A.N.; Hass, C.J.; Nichols, T.R.; Gregor, R.J. The effects of sloped surfaces on locomotion: An electromyographic analysis. *J. Biomech.* **2007**, *40*, 1276–1285. [CrossRef] [PubMed]
45. Alexander, N.; Schwameder, H. Effect of sloped walking on lower limb muscle forces. *Gait Posture* **2016**, *47*, 62–67. [CrossRef] [PubMed]
46. Ma, L.; Yang, Y.; Chen, N.; Song, R.; Li, L. Effect of different terrains on onset timing, duration and amplitude of tibialis anterior activation. *Biomed. Signal. Process. Control.* **2015**, *19*, 115–121. [CrossRef]
47. Li, B.; Gui, Q.; Ali, H.B.; Li, H.; Jin, Z. A wearable sit-to-stand detection system based on angle tracking and lower limb EMG. In Proceedings of the 2016 IEEE Signal Processing in Medicine and Biology Symposium (SPMB), Philadelphia, PA, USA, 3 December 2016; pp. 1–6.
48. Pappas, I.P.; Popovic, M.R.; Keller, T.; Dietz, V.; Morari, M. A reliable gait phase detection system. *IEEE Trans. Neural Syst. Rehabil. Eng.* **2001**, *9*, 113–125. [CrossRef]
49. Perry, J.; Davids, J.R. Gait analysis: Normal and pathological function. *J. Pediatr. Orthop.* **1992**, *12*, 815. [CrossRef]
50. Jung, J.Y.; Heo, W.; Yang, H.; Park, H. A Neural Network-Based Gait Phase Classification Method using Sensors Equipped on Lower Limb Exoskeleton Robots. *Sensors* **2015**, *15*, 27738–27759. [CrossRef]
51. Abdulhay, E.; Arunkumar, N.; Narasimhan, K.; Vellaiappan, E.; Venkatraman, V. Gait and tremor investigation using machine learning techniques for the diagnosis of Parkinson disease. *Future Gener. Comput. Syst.* **2018**, *83*, 366–373. [CrossRef]
52. Simão, M.; Mendes, N.; Gibaru, O.; Neto, P. A review on electromyography decoding and pattern recognition for human-machine interaction. *IEEE Access* **2019**, *7*, 39564–39582. [CrossRef]
53. Smith, L.H.; Hargrove, L.J. Comparison of surface and intramuscular EMG pattern recognition for simultaneous wrist/hand motion classification. In Proceedings of the 2013 35th annual international conference of the IEEE engineering in medicine and biology society (EMBC), Osaka, Japan, 3–7 July 2013; pp. 4223–4226.
54. Smith, L.H.; Kuiken, T.A.; Hargrove, L.J. Real-time simultaneous and proportional myoelectric control using intramuscular EMG. *J. Neural Eng.* **2014**, *11*, 066013. [CrossRef] [PubMed]
55. Leonard, C.T.; Brown, J.S.; Price, T.R.; Queen, S.A.; Mikhailenok, E.L. Comparison of surface electromyography and myotonometric measurements during voluntary isometric contractions. *J. Electromyogr. Kinesiol.* **2004**, *14*, 709–714. [CrossRef] [PubMed]
56. Samadani, A.-A.; Kulic, D. Hand gesture recognition based on surface electromyography. In Proceedings of the 2014 36th Annual International Conference of the IEEE Engineering in Medicine and Biology Society, Chicago, IL, USA, 26–30 August 2014; pp. 4196–4199.
57. Du, Y.-C.; Lin, C.-H.; Shyu, L.-Y.; Chen, T. Portable hand motion classifier for multi-channel surface electromyography recognition using grey relational analysis. *Expert Syst. Appl.* **2010**, *37*, 4283–4291. [CrossRef]
58. Farrell, T.R. A comparison of the effects of electrode implantation and targeting on pattern classification accuracy for prosthesis control. *IEEE Trans. Biomed. Eng.* **2008**, *55*, 2198–2211. [CrossRef]
59. Martelloni, C.; Carpaneto, J.; Micera, S. Classification of upper arm EMG signals during object-specific grasp. In Proceedings of the Conference proceedings: Annual International Conference of the IEEE Engineering in Medicine and Biology Society, Vancouver, BC, Canada, 21–24 August 2008; pp. 5061–5064.
60. Fang, C.; Wang, Y.; Gao, S. Flexible and Wearable GRF and EMG Sensors Enabled Locomotion Mode Recognition for IoHT Based In-Home Rehabilitation. Available online: https://www.techrxiv.org/articles/Flexible_and_Wearable_GRF_and_EMG_Sensors_Enabled_Locomotion_Mode_Recognition_for_IoHT_Based_In-home_Rehabilitation/11987994/1 (accessed on 21 March 2020).
61. Li, L.; Ogden, L.L. Muscular activity characteristics associated with preparation for gait transition. *J. Sport Health Sci.* **2012**, *1*, 27–35. [CrossRef]
62. Yousif, A.E.; Aziz, M.Y. Biomechanical analysis of the human femur bone during going upstairs and sitting down. In Proceedings of the 2012 First National Conference for Engineering Sciences (FNCES 2012), Baghdad, Iraq, 7–8 November 2012; pp. 1–7.

63. Pati, S.; Joshi, D.; Mishra, A. Locomotion classification using EMG signal. In Proceedings of the 2010 International Conference on Information and Emerging Technologies, Karachi, Pakistan, 14–16 June 2010; pp. 1–6.
64. Chiu, M.-C.; Wang, M.-J. The effect of gait speed and gender on perceived exertion, muscle activity, joint motion of lower extremity, ground reaction force and heart rate during normal walking. *Gait Posture* **2007**, *25*, 385–392. [CrossRef]
65. Komi, P.V.; Tesch, P. EMG frequency spectrum, muscle structure, and fatigue during dynamic contractions in man. *Eur. J. Appl. Physiol. Occup. Physiol.* **1979**, *42*, 41–50. [CrossRef]
66. Larsson, B.; Månsson, B.; Karlberg, C.; Syvertsson, P.; Elert, J.; Gerdle, B. Reproducibility of surface EMG variables and peak torque during three sets of ten dynamic contractions. *J. Electromyogr. Kinesiol.* **1999**, *9*, 351–357. [CrossRef]
67. Wang, J.; Tang, L.; Bronlund, J.E. Surface EMG signal amplification and filtering. *Int. J. Comput. Appl.* **2013**, *82*, 15–22. [CrossRef]
68. Barzilay, O.; Wolf, A. A fast implementation for EMG signal linear envelope computation. *J. Electromyogr. Kinesiol.* **2011**, *21*, 678–682. [CrossRef]
69. Dutta, A.; Khattar, B.; Banerjee, A. Nonlinear Analysis of Electromyogram Following Neuromuscular Electrical Stimulation-Assisted Gait Training in Stroke Survivors. In *Converging Clinical and Engineering Research on Neurorehabilitation*; Springer: Berlin/Heidelberg, Germany, 2013; pp. 53–57.
70. Barsakcioglu, D.Y.; Farina, D. A real-time surface emg decomposition system for non-invasive human-machine interfaces. In Proceedings of the 2018 IEEE Biomedical Circuits and Systems Conference (BioCAS), Cleveland, OI, USA, 17–19 October 2018; pp. 1–4.
71. Hudgins, B.; Parker, P.; Scott, R.N. A new strategy for multifunction myoelectric control. *IEEE Trans. Biomed. Eng.* **1993**, *40*, 82–94. [CrossRef] [PubMed]
72. Englehart, K.; Hudgins, B. A robust, real-time control scheme for multifunction myoelectric control. *IEEE Trans. Biomed. Eng.* **2003**, *50*, 848–854. [CrossRef] [PubMed]
73. Zardoshti-Kermani, M.; Wheeler, B.C.; Badie, K.; Hashemi, R.M. EMG feature evaluation for movement control of upper extremity prostheses. *IEEE Trans. Rehabil. Eng.* **1995**, *3*, 324–333. [CrossRef]
74. Englehart, K.; Hugdins, B.; Parker, P. *Multifunction Control of Prostheses Using the Myoelectric Signal*; CRC Press: New York, NY, USA, 2000; pp. 153–208.
75. Afzal, T.; Iqbal, K.; White, G.; Wright, A.B. A method for locomotion mode identification using muscle synergies. *IEEE Trans. Neural Syst. Rehabil. Eng.* **2016**, *25*, 608–617. [CrossRef]
76. Phinyomark, A.; Phukpattaranont, P.; Limsakul, C. Feature reduction and selection for EMG signal classification. *Expert Syst. Appl.* **2012**, *39*, 7420–7431. [CrossRef]
77. Englehart, K.; Hudgins, B.; Parker, P.A.; Stevenson, M. Classification of the myoelectric signal using time-frequency based representations. *Med. Eng. Phys.* **1999**, *21*, 431–438. [CrossRef]
78. Samuel, O.W.; Zhou, H.; Li, X.; Wang, H.; Zhang, H.; Sangaiah, A.K.; Li, G. Pattern recognition of electromyography signals based on novel time domain features for amputees' limb motion classification. *Comput. Electr. Eng.* **2018**, *67*, 646–655. [CrossRef]
79. Hargrove, L.J.; Li, G.; Englehart, K.B.; Hudgins, B.S. Principal components analysis preprocessing for improved classification accuracies in pattern-recognition-based myoelectric control. *IEEE Trans. Biomed. Eng.* **2008**, *56*, 1407–1414. [CrossRef]
80. Naik, G.R.; Nguyen, H.T. Nonnegative matrix factorization for the identification of EMG finger movements: Evaluation using matrix analysis. *IEEE J. Biomed. Health Inform.* **2014**, *19*, 478–485. [CrossRef]
81. Naik, G.R.; Kumar, D.K.; Palaniswami, M. Surface EMG based hand gesture identification using semi blind ICA: Validation of ICA matrix analysis. *Electromyogr. Clin. Neurophysiol.* **2008**, *48*, 169–180.
82. Gallant, P.; Morin, E.; Peppard, L. Feature-based classification of myoelectric signals using artificial neural networks. *Med. Biol. Eng. Comput.* **1998**, *36*, 485–489. [CrossRef] [PubMed]
83. Naik, G.R.; Selvan, S.E.; Gobbo, M.; Acharyya, A.; Nguyen, H.T. Principal component analysis applied to surface electromyography: A comprehensive review. *IEEE Access* **2016**, *4*, 4025–4037. [CrossRef]
84. Chen, M.; Zhou, P. A novel framework based on FastICA for high density surface EMG decomposition. *IEEE Trans. Neural Syst. Rehabil. Eng.* **2015**, *24*, 117–127. [CrossRef] [PubMed]
85. Lee, D.D.; Seung, H.S. Algorithms for non-negative matrix factorization. In Proceedings of the Advances in neural information processing systems, Vancouver, BC, Canada, 3–8 December 2001; pp. 556–562.

86. Farina, D.; Jiang, N.; Rehbaum, H.; Holobar, A.; Graimann, B.; Dietl, H.; Aszmann, O.C. The extraction of neural information from the surface EMG for the control of upper-limb prostheses: Emerging avenues and challenges. *IEEE Trans. Neural Syst. Rehabil. Eng.* **2014**, *22*, 797–809. [CrossRef]
87. Liu, J.; Zhou, P. A novel myoelectric pattern recognition strategy for hand function restoration after incomplete cervical spinal cord injury. *IEEE Trans. Neural Syst. Rehabil. Eng.* **2012**, *21*, 96–103. [CrossRef]
88. Zhou, P.; Lowery, M.M.; Englehart, K.B.; Huang, H.; Li, G.; Hargrove, L.; Dewald, J.P.; Kuiken, T.A. Decoding a new neural–machine interface for control of artificial limbs. *J. Neurophysiol.* **2007**, *98*, 2974–2982. [CrossRef]
89. Nazarpour, K.; Sharafat, A.; Firoozabadi, S. Surface EMG signal classification using a selective mix of higher order statistics. In Proceedings of the 2005 IEEE Engineering in Medicine and Biology 27th Annual Conference, Shanghai, China, 1–4 September 2005; pp. 4208–4211.
90. Chowdhury, R.H.; Reaz, M.B.I.; Ali, M.A.B.; Bakar, A.A.A.; Chellappan, K.; Chang, T.G. Surface Electromyography Signal Processing and Classification Techniques. *Sensors* **2013**, *13*, 12431–12466. [CrossRef]
91. Al-Angari, H.M.; Kanitz, G.; Tarantino, S.; Cipriani, C. Distance and mutual information methods for EMG feature and channel subset selection for classification of hand movements. *Biomed. Signal. Process. Control.* **2016**, *27*, 24–31. [CrossRef]
92. Boostani, R.; Moradi, M.H. Evaluation of the forearm EMG signal features for the control of a prosthetic hand. *Physiol. Meas.* **2003**, *24*, 309. [CrossRef]
93. Too, J.; Abdullah, A.R.; Mohd Saad, N.; Tee, W. Emg feature selection and classification using a pbest-guide binary particle swarm optimization. *Computation* **2019**, *7*, 12. [CrossRef]
94. Phinyomark, A.; Hirunviriya, S.; Limsakul, C.; Phukpattaranont, P. Evaluation of EMG feature extraction for hand movement recognition based on Euclidean distance and standard deviation. In Proceedings of the ECTI-CON2010: The 2010 ECTI International Confernce on Electrical Engineering/Electronics, Computer, Telecommunications and Information Technology, Chiang Mai, Thailand, 19–21 May 2010; pp. 856–860.
95. Huang, H.; Xie, H.-B.; Guo, J.-Y.; Chen, H.-J. Ant colony optimization-based feature selection method for surface electromyography signals classification. *Comput. Biol. Med.* **2012**, *42*, 30–38. [CrossRef] [PubMed]
96. Liu, J.; Li, X.; Li, G.; Zhou, P. EMG feature assessment for myoelectric pattern recognition and channel selection: A study with incomplete spinal cord injury. *Med. Eng. Phys.* **2014**, *36*, 975–980. [CrossRef] [PubMed]
97. Benazzouz, A.; Guilal, R.; Amirouche, F.; Slimane, Z.E.H. EMG Feature Selection for Diagnosis of Neuromuscular Disorders. In Proceedings of the 2019 International Conference on Networking and Advanced Systems (ICNAS), Annaba, Algeria, 26–27 June 2019; pp. 1–5.
98. Gupta, R.; Agarwal, R. Single channel EMG-based continuous terrain identification with simple classifier for lower limb prosthesis. *Biocybern. Biomed. Eng.* **2019**, *39*, 775–788. [CrossRef]
99. Pires, R.; Falcari, T.; Campo, A.B.; Pulcineli, B.C.; Hamill, J.; Ervilha, U.F. Using a Support Vector Machine Algorithm to Classify Lower-Extremity EMG Signals During Running Shod/Unshod With Different Foot Strike Patterns. *J. Appl. Biomech.* **2019**, *35*, 87–90. [CrossRef]
100. Li, H.-T.; Han, S.-L.; Pan, M.-C. Lower-limb motion classification for hemiparetic patients through IMU and EMG signal processing. In Proceedings of the 2016 International Conference on Biomedical Engineering (BME-HUST), Hanoi, Vietnam, 4–6 October 2016; pp. 113–118.
101. Mesbah, S.; Gonnelli, F.; El-baz, A.; Angeli, C.; Harkema, S.; Rejc, E. Spectral analysis of lower limb EMG activity in individuals with motor complete SCI during standing with epidural stimulation. In Proceedings of the 2018 IEEE International Symposium on Signal Processing and Information Technology (ISSPIT), Louisville, KY, USA, 6–8 December 2018; pp. 1–5.
102. Prasad, C.; Balakandan, V.K.; Moorthy, P.; Kochuvila, S. Classification of sEMG Signals for Controlling of a Prosthetic foot using SVM and KNN. In Proceedings of the 2019 International Conference on Intelligent Computing and Control Systems (ICCS), Madurai, India, 15–17 May 2019; pp. 454–458.
103. Xie, H.-B.; Guo, T.; Bai, S.; Dokos, S. Hybrid soft computing systems for electromyographic signals analysis: A review. *Biomed. Eng. Online* **2014**, *13*, 8. [CrossRef]
104. Meng, M.; Luo, Z.; She, Q.; Ma, Y. Automatic recognition of gait mode from EMG signals of lower limb. In Proceedings of the 2010 The 2nd International Conference on Industrial Mechatronics and Automation, HongKong, China, 30 May 2010; pp. 282–285.

105. Morbidoni, C.; Principi, L.; Mascia, G.; Strazza, A.; Verdini, F.; Cucchiarelli, A.; Di Nardo, F. Gait phase classification from surface EMG signals using Neural Networks. In *Mediterranean Conference on Medical and Biological Engineering and Computing*; Springer: Cham, Switzerland, 2019; pp. 75–82.
106. Lendaro, E.; Mastinu, E.; Håkansson, B.; Ortiz-Catalan, M. Real-time classification of non-weight bearing lower-limb movements using EMG to facilitate phantom motor execution: Engineering and case study application on phantom limb pain. *Front. Neurol.* **2017**, *8*, 470. [CrossRef]
107. Zheng, J.; Chen, J.; Yang, M.; Chen, S. PSO-SVM-based gait phase classification during human walking on unstructured terrains: Application in lower-limb exoskeleton. *Proc. Inst. Mech. Eng. Part. C J. Mech. Eng. Sci.* **2019**, *233*, 7144–7154. [CrossRef]
108. Bellingegni, A.D.; Gruppioni, E.; Colazzo, G.; Davalli, A.; Sacchetti, R.; Guglielmelli, E.; Zollo, L. NLR, MLP, SVM, and LDA: A comparative analysis on EMG data from people with trans-radial amputation. *J. Neuroeng. Rehabil.* **2017**, *14*, 82. [CrossRef]
109. Chen, J.; Zhang, X.; Cheng, Y.; Xi, N. Surface EMG based continuous estimation of human lower limb joint angles by using deep belief networks. *Biomed. Signal. Process. Control.* **2018**, *40*, 335–342. [CrossRef]
110. Afzal, T.; White, G.; Wright, A.B.; Iqbal, K. Locomotion mode identification for lower limbs using neuromuscular and joint kinematic signals. In Proceedings of the 2014 36th Annual International Conference of the IEEE Engineering in Medicine and Biology Society, Chicago, IL, USA, 27–31 August 2014; pp. 4071–4074.
111. Blana, D.; Kyriacou, T.; Lambrecht, J.M.; Chadwick, E.K. Feasibility of using combined EMG and kinematic signals for prosthesis control: A simulation study using a virtual reality environment. *J. Electromyogr. Kinesiol* **2016**, *29*, 21–27. [CrossRef]
112. Kundu, A.S.; Mazumder, O.; Lenka, P.K.; Bhaumik, S. Hand gesture recognition based omnidirectional wheelchair control using IMU and EMG sensors. *J. Intell. Robot. Syst.* **2018**, *91*, 529–541. [CrossRef]
113. Ceolini, E.; Taverni, G.; Khacef, L.; Payvand, M.; Donati, E. Sensor fusion using EMG and vision for hand gesture classification in mobile applications. In Proceedings of the 2019 IEEE Biomedical Circuits and Systems Conference (BioCAS), Nara, Japan, 17–19 October 2019; pp. 1–4.
114. Pedotti, A.; Assente, R.; Fusi, G.; De Rossi, D.; Dario, P.; Domenici, C. Multisensor piezoelectric polymer insole for pedobarography. *Ferroelectrics* **1984**, *60*, 163–174. [CrossRef]
115. Hidler, J. Robotic-assessment of walking in individuals with gait disorders. In Proceedings of the 26th Annual International Conference of the IEEE Engineering in Medicine and Biology Society, San Francisco, CA, USA, 1–4 September 2004; pp. 4829–4831.
116. Riener, R.; Rabuffetti, M.; Frigo, C.; Quintern, J.; Schmidt, G. Instrumented staircase for ground reaction measurement. *Med. Biol. Eng. Comput.* **1999**, *37*, 526–529. [CrossRef] [PubMed]
117. Faivre, A.; Dahan, M.; Parratte, B.; Monnier, G. Instrumented shoes for pathological gait assessment. *Mech. Res. Commun.* **2004**, *31*, 627–632. [CrossRef]
118. Lacković, I.; Bilas, V.; Šantić, A.; Nikolić, V. Measurement of gait parameters from free moving subjects. *Measurement* **2000**, *27*, 121–131. [CrossRef]
119. Šantić, A.; Bilas, V.; Lacković, I. A system for measurement forces in feet and crutches during normal na pathological gait. *Period. Biol.* **2002**, *104*, 305–310.
120. Hennig, E.M.; Staats, A.; Rosenbaum, D. Plantar pressure distribution patterns of young school children in comparison to adults. *Foot Ankle Int.* **1994**, *15*, 35–40. [CrossRef]
121. Zadravec, M.; Olenšek, A.; Matjačić, Z. Emulation of hill walking and turning on Balance Assessment Robot: A preliminary study. In Proceedings of the 2019 IEEE 16th International Conference on Rehabilitation Robotics (ICORR), Toronto, ON, Canada, 24–28 June 2019; pp. 7–12.
122. KYEONG, S.; SHIN, W.; YANG, M.; HEO, U.; FENG, J.-r.; KIM, J. Recognition of walking environments and gait period by surface electromyography. *Front. Inf. Technol. Electron. Eng.* **2019**, *20*, 342–352. [CrossRef]
123. Galli, M.; Cimolin, V.; Crugnola, V.; Priano, L.; Menegoni, F.; Trotti, C.; Milano, E.; Mauro, A. Gait pattern in myotonic dystrophy (Steinert disease): A kinematic, kinetic and EMG evaluation using 3D gait analysis. *J. Neurol. Sci.* **2012**, *314*, 83–87. [CrossRef]
124. Cipriani, C.; Antfolk, C.; Controzzi, M.; Lundborg, G.; Rosén, B.; Carrozza, M.C.; Sebelius, F. Online myoelectric control of a dexterous hand prosthesis by transradial amputees. *IEEE Trans. Neural Syst. Rehabil. Eng.* **2011**, *19*, 260–270. [CrossRef]

125. Young, A.; Hargrove, L. Strategies to reduce myoelectric pattern recognition sensitivity to electrode shift. In Proceedings of the 18th Congress of the International Society of Electrophysiology and Kinesiology, Aalborg, Denmark, 17–19 June 2010; pp. 16–19.
126. Geng, W.; Du, Y.; Jin, W.; Wei, W.; Hu, Y.; Li, J. Gesture recognition by instantaneous surface EMG images. *Sci. Rep.* **2016**, *6*, 36571. [CrossRef] [PubMed]
127. Amma, C.; Krings, T.; Böer, J.; Schultz, T. Advancing muscle-computer interfaces with high-density electromyography. In Proceedings of the 33rd Annual ACM Conference on Human Factors in Computing Systems, Seoul, Korea, 18–23 April 2015; pp. 929–938.
128. Winter, D.A.; Yack, H. EMG profiles during normal human walking: Stride-to-stride and inter-subject variability. *Electroencephalogr. Clin. Neurophysiol.* **1987**, *67*, 402–411. [CrossRef]
129. Phinyomark, A.; Scheme, E. EMG pattern recognition in the era of big data and deep learning. *Big Data Cogn. Comput.* **2018**, *2*, 21. [CrossRef]
130. Gorgolewski, K.; Margulies, D.S.; Milham, M.P. Making data sharing count: A publication-based solution. *Front. Neurosci.* **2013**, *7*, 9. [CrossRef]
131. Atzori, M.; Gijsberts, A.; Heynen, S.; Hager, A.-G.M.; Deriaz, O.; Van Der Smagt, P.; Castellini, C.; Caputo, B.; Müller, H. Building the Ninapro database: A resource for the biorobotics community. In Proceedings of the 2012 4th IEEE RAS & EMBS International Conference on Biomedical Robotics and Biomechatronics (BioRob), Rome, Italy, 24–27 June 2012; pp. 1258–1265.
132. Young, A.J.; Simon, A.M.; Hargrove, L.J. A Training Method for Locomotion Mode Prediction Using Powered Lower Limb Prostheses. *IEEE Trans. Neural Syst. Rehabil. Eng.* **2014**, *22*, 671–677. [CrossRef]
133. Martinez-Hernandez, U.; Rubio-Solis, A.; Panoutsos, G.; Dehghani-Sanij, A.A. A combined adaptive neuro-fuzzy and bayesian strategy for recognition and prediction of gait events using wearable sensors. In Proceedings of the 2017 IEEE International Conference on Fuzzy Systems (FUZZ-IEEE), Naples, Italy, 9–12 July 2017; pp. 1–6.
134. Blana, D.; Van Den Bogert, A.J.; Murray, W.M.; Ganguly, A.; Krasoulis, A.; Nazarpour, K.; Chadwick, E.K. Model-based control of individual finger movements for prosthetic hand function. *IEEE Trans. Neural Syst. Rehabil. Eng.* **2020**, *28*, 612–620. [CrossRef]

MDPI
St. Alban-Anlage 66
4052 Basel
Switzerland
Tel. +41 61 683 77 34
Fax +41 61 302 89 18
www.mdpi.com

Biosensors Editorial Office
E-mail: biosensors@mdpi.com
www.mdpi.com/journal/biosensors

www.ingramcontent.com/pod-product-compliance
Lightning Source LLC
Chambersburg PA
CBHW062101220926
51850CB00010B/2855
* 9 7 8 3 0 3 6 5 2 4 5 4 2 *